W9-BNC-614

Galileo's Mistake

ALSO BY WADE ROWLAND

Ockham's Razor: A Search for Wonder in an Age of Doubt

Spirit of the Web: The Age of Information from Telegraph to Internet

Galileo's Mistake

A New Look at the Epic Confrontation between Galileo and the Church

Wade Rowland

Arcade Publishing • New York

FIRST U.S. EDITION 2003

First published in Canada in 2001 by Thomas Allen Publishers and revised for this edition

Library of Congress Cataloging-in-Publication Data

Rowland, Wade.
Galileo's mistake: a new look at the epic confrontation between Galileo and the Church / Wade Rowland. —1st U.S. ed.
p. cm.
Includes bibliographical references and index.
ISBN 1-55970-684-8
1. Galilei, Galileo, 1564–1642. 2. Galilei, Galileo, 1564–1642—Trials, litigation, etc. 3. Religion and science—Italy—History—17th century. 4. Science—Philosophy. I. Title.

QB36.G2R62 2003
520'.92—dc21 2002044054

Published in the United States by Arcade Publishing, Inc., New York
Distributed by AOL Time Warner Book Group

Visit our Web site at www.arcadepub.com
Visit the author's Web site at www.galileosmistake.com

10 9 8 7 6 5 4 3 2 1

EB

PRINTED IN THE UNITED STATES OF AMERICA

To the memory of my mother,

Edith Mary Rowland

Humanity has in course of time had to endure from the hand of science two great outrages upon its naive self-love. The first was when it realized that our Earth was not the center of the universe, but only a speck in a world system of a magnitude hardly conceivable . . . the second was when biological research robbed man of his particular privilege of having been specially created and relegated him to a descent from the animal world.

—SIGMUND FREUD

We are entitled to require a consistency between what people write in their studies and the way in which they live their lives. I submit that no one lives as if science were enough. Our account of the world must be rich enough—have a thick enough texture and a sufficiently generous rationality—to contain the total spectrum of human meeting with reality.

—JOHN POLKINGHORNE

CONTENTS

ILLUSTRATION AND PHOTO CREDITS

Every reasonable effort has been made to contact the holders of copyright for materials reproduced in this book. The publisher will gladly receive any information about errors or omissions herein to be corrected in subsequent editions of the book.

Photo insert

Galileo in middle age, from the Toronto Metro Reference Library Picture Collection, artist unknown.

The Council of Trent, reproduction of the image from collection of the author. Artist unknown, Basilica of Santa Maria in Trastavere, Rome.

Johannes Kepler, from http://www.groups.dcs.standrews.ac.uk/~history/PictDisplay/Kepler.html

Nicolaus Copernicus, artist unknown, from http://www.er.uqam.ca/nobel/r14310/Ptolemy/Copernic/index.html

Galileo's sketches of the Moon, SCALA/Art Resource.

Galileo's sketch of sunspots, from *The Cambridge Companion to Galileo,* Peter Machamer, ed., Cambridge University Press, 1998.

Ceiling fresco at Santa Maria Maggiore, from *Il Cigoli* by Roberto Contini, Edizioni dei Soncino, 1991.

Camilo Borghese (Pope Paul V), Hulton Archive/Stone.

Pope Urban VIII by Pietro da Cortona, Capitoline Museum, Rome. Reproduction of image from collection of the author.

Galileo before the Inquisitional Tribunal by Cristiano Banti. Photo by Foto Sapporetti, from *Galileo Heretic* by Pietro Redondi, Princeton University Press, Princeton, NJ, 1987.

Galileo's tomb, from *Santa Croce: The Church, The Cloisters and The Museum,* English ed., Mario Franchi Becocci. No publication date available.

Diagrams in text

Dante's world view, from *A Short History of Scientific Ideas to 1900* by Charles Singer, Oxford University Press. Reprinted with permission of Oxford University Press.

Ptolemy's construction of planetary motion, from *A Short History of Scientific Ideas to 1900* by Charles Singer, Oxford University Press. Reprinted with permission of Oxford University Press.

Copernicus' descriptions of planetary orbits, from *The Copernican Revolution: Planetary Astronomy in the Development of Western Thought* by Thomas S. Kuhn, Cambridge, MA: Harvard University Press, copyright © 1957 by the President and Fellows of Harvard College, copyright © 1985 by Thomas S. Kuhn. Reprinted with permission of Harvard University Press.

The annual stellar parallax, from *The Copernican Revolution: Planetary Astronomy in the Development of Western Thought* by Thomas S. Kuhn, Cambridge, MA: Harvard University Press, copyright © 1957 by the President and Fellows of Harvard College, copyright © 1985 by Thomas S. Kuhn. Reprinted with permission of Harvard University Press.

Galileo's log of Jupiter's moons, from *The Cambridge Companion to Galileo,* Peter Machamer, ed., Cambridge University Press, 1998. Copyright holder unknown.

ACKNOWLEDGMENTS

I am grateful for the criticism and encouragement of members of the faculty of the graduate program in Methodologies for the Study of Western History and Culture at Trent University in Ontario, especially Professors Andrew Wernick, David Holdsworth, and B. J. Hodgson, and Professor C. V. Boundas, chairman of the university's department of philosophy. Galileo scholar Michael John Gorman was kind enough to share his insights over a wonderful dinner in Rome and in subsequent correspondence.

Thanks to Douglas Rowland, Bruce Powe, Michael Holmes, and Derrick De Kerckhove for reading the manuscript and offering valuable criticism. Alison Reid provided a remarkably thorough copy-edit that improved the text in many ways, and Katja Pantzar gave careful stewardship of the production process. Casey Ebro gave me much thoughtful editing advice for this American edition.

Roberto Benardi and Terry Storms of the Canadian Embassy to the Holy See provided invaluable assistance in arranging various permissions with the Vatican and the Italian government. Caterina Fasella was my very knowledgeable guide through the convent of Santa Maria sopre Minerva, where Galileo was tried by the Inquisition. My son, Simon, gamely trudged many miles with me through the streets and alleys of Rome in search of Galileo's world.

Relais et Châteaux in Paris offered their usual exemplary services in arranging accommodation throughout Tuscany during my initial research trip to Italy. My thanks in particular to Maryse Masse and Régis Bulot.

I am deeply indebted to Patrick Crean for his continuing support and enthusiasm for my literary projects.

My wife, Christine Collie Rowland, makes all things possible for me. No finer travel companion could be imagined.

Galileo's Mistake

ONE

Overture

The story of the astronomer and mathematician Galileo Galilei and his trial for heresy by the Inquisition is one of the defining narratives of modern Western culture. The moral lessons it teaches are a cornerstone of our belief in the supreme power and validity of reason, and in science's exclusive access to reliable knowledge of the world we live in. It is a tale that vividly illustrates the dangers and arbitrariness of religious authority, and the futility of resistance to the inexorable advance of scientific knowledge.

There is a modest historical marker in Rome, outside the magnificent Villa Medici where Galileo stayed during his visits to that city, and it sums up what might be called the authorized version of the story. I discovered it one brilliant morning in May in the first year of the new millennium, where the enclosing villa wall was softened by masses of mauve delphiniums just coming into bloom. The monument was placed here in 1887. It is about ten feet tall overall, half its height a cylinder of greenish marble, capped with a white mushroom dome on a white fluted marble plinth. A heavy bronze sleeve is wrapped around the middle of the marble cylinder, and on this is an inscription. It says, "*It was here that Galileo was kept prisoner by the Holy Office, when he was on trial for having seen that the Earth moves and the Sun stands still.*"*

---

*The Congregation of the Holy Office was inaugurated in 1542, at the height of the Counter-Reformation, by Pope Paul III. It was specifically instructed to take

The marker is ignored in the voluminous travel literature on Rome—no guidebook that I have seen so much as mentions it, although most identify the villa itself as the place where Galileo was confined while being tried by the Inquisition. One might easily conclude that the lack of attention paid the monument is due to the fact that it merely identifies a geographical feature and is of no more intrinsic interest than a street sign. But it is a cultural artifact of real significance. The reason is that it expresses one of the central misconceptions of the authorized version of the Galileo story—what might more properly be called the myth of Galileo. And that is that he was condemned by the Catholic Church for having discovered the truth. It was the Galileo scholar Maurice Finocchiaro who led me to the marker, in an article in which he asserts that since "to condemn a person for such a reason [that is, for having discovered the truth] can only be the result of ignorance and narrow-mindedness, this is also the myth which is used to justify the incompatibility between science and religion."[1] For Finocchiaro and most other current historians and philosophers of science, the myth is erroneous, simplistic, and misleading.

Nevertheless, it is so widespread that a version of it, not much more sophisticated than that displayed on the monument, is presented in Albert Einstein's introduction to the standard English translation of *Dialogue on the Two Chief World Systems: Ptolemaic and Copernican*, the work for which Galileo was condemned. And the myth is dramatized to sensational effect in *Life of Galileo*, Bertolt Brecht's seductively brilliant play from 1938–39, made into a film in 1975 by the American director Joseph Losey. Children in grade school are asked to write essays on Galileo as a martyr to intellectual freedom. And the Church itself has in recent years seemed to lend credence to the myth with its own repentant attitude.

My own suspicion that there might be more to the story of Galileo than is contained in the orthodox version was stirred long ago by

---

over the suppression of heresies and heretics, formerly the purview of the Inquisition. From that time forward, the names Inquisition and Holy Office became almost synonymous. At the Second Vatican Council in 1965 the name of the Holy Office was changed to the Congregation for the Doctrine of the Faith.

questions that haunted the blood-drenched twentieth century. How could a civilization that generated the technical marvels that ease our lives in so many ways also have spawned fascism, genocides, environmental havoc, weapons of mass destruction? Why was the once-brilliant promise of "the good life" an ever-receding chimera? A phenomenon of our time has been the rise of Dickensian mean-spiritedness as the foundation for a respectable social credo. Our worst-paid and least prestigious jobs are those that involve benevolence, as in caring for the disabled, the disturbed, the old, and the chronically sick. For most people, in most countries, increasing workloads have meant the virtual abandonment of family life, not to mention civic responsibilities. Quality of life has seemed to change in inverse ratio to leading economic indicators. Why are we, in spite of our announced humanist intentions, increasingly treating our fellow human beings as means to economic ends rather than as ends in themselves? How did we come to justify treating government as a purely economic entity, subject to the crude accounting practices of business?

While some philosophers have found answers to these puzzling questions in the rationalism that swept the West in the eighteenth century, and others blamed the romanticism and anti-rationalism that followed in reaction, neither of these views seem adequate to me. I came to share a conviction that the roots of what is most disturbing in the modern world find their nourishment deeper in history, in what is often called the Scientific Revolution. This revolution began in seventeenth-century Europe, and Galileo was among its earliest instigators. Dramatic advances were made in several fields of inquiry—notably mathematics, physics, and astronomy. Pascal invented the calculating machine, Leibniz and Newton jointly invented calculus, Robert Boyle laid the foundations of modern chemistry, and William Harvey mapped the body's circulatory system. John Napier eased the enormous burden of astronomical calculations with his invention of logarithms. Descartes introduced his Cartesian coordinates and gave geometry a new dimension. The microscope, like the telescope, accurate pendulum clocks, and balance-wheel watches, came into general use. Galileo made his epochal discoveries in mechanics and astronomy.

At the same time, radical new ideas were abroad about the

nature of knowledge and how best to go about acquiring it, and once again Galileo was in the forefront. The one endeavor reinforced the other, so that these revolutionary insights into the workings of the world and how to explore it had an enormous impact on philosophy, and the names of the great scientists are linked with those of philosophers like Thomas Hobbes and John Locke and Francis Bacon. Prodigies like Galileo and Descartes and Leibniz excelled in both science and philosophy—indeed, the fields were for much of the century thought of as related aspects of the same discipline. Theology, too, felt the Scientific Revolution's impact—in fact, was staggered by it. Beginning in the late seventeenth and throughout the eighteenth century, the Church of Rome, for more than fifteen hundred years a leading political force in Europe, was stripped of most of its secular power and then of much of its once-universal moral authority as well. The effect on an institution already suffering from the hammer blows of the Protestant Reformation and the chaos of the Thirty Years War was ultimately devastating.

We are still, with increasing unease, living with the results of that historic shift in outlook and values. On the one hand, the Scientific Revolution endowed Western civilization with the ability to manipulate nature to an almost magical degree. On the other, it prompted a shift in the prevailing view of the acquisition of knowledge and of moral thought that deprived civilization of any effective means to manage the career of science and to ameliorate its unwanted impacts. It bequeathed unprecedented power and wealth while at the same time undermining the foundations of the wisdom necessary to their judicious and benevolent use. It expanded the creative horizons of humanity while reducing the mass of individual humans to the status of commodities and consumers. It improved health and longevity while promoting unprecedented spiritual and existential dis-ease.

Amid all the political turmoil and intellectual ferment of this watershed period, it seemed to me that the Church of Rome's epic confrontation with Galileo was a supremely significant event, one that presents in microcosm the issues that define this most portentous turning point of the second millennium, the transition from the Age of Faith to the Age of Reason—from an era of religion and spirituality to an epoch of science and materialism. Understanding that seminal episode in the history of the modern world can, I believe, provide valuable insights into many of the most vexatious problems afflicting contemporary life, and, more important, clues to finding solutions.

Unfortunately, it is clear from the most cursory examination of school texts, popular literature, science journalism, and even academic treatises that although the historical significance of Galileo's trial is widely conceded, the nature of that significance is almost universally misunderstood. The popular conception of the confrontation and the trial that was its culmination has, indeed, changed little in the more than 350 years since it took place. The Church's victory at the trial with Galileo's conviction was a Pyrrhic one, and the scientist's controversial ideas won an overwhelming triumph in the wider war. It is the victors who write the history of wars, and so it was the heroic picture of Galileo as a lonely champion of enlightenment and the Church as a blind, despotic power, hostile to scientific inquiry, that has come down to us.

If the familiar myth of Galileo as the paladin of truth and freedom opposing a venal and closed-minded Church is untrue, as I have asserted, what really happened between Galileo and the Church back in the seventeenth century? A central issue in the events surrounding the trial was the Copernican hypothesis, the idea that the Earth moved with both diurnal rotations and annual revolutions around the Sun. Galileo's supposed "heresy" lay—at least nominally—in advocating Copernicanism in the face of apparently contradictory biblical passages. The hypothesis, as everybody knows, is correct. What is not so widely known, however, is that there was no convincing proof of its correctness in Galileo's time. Even less well known is the fact that despite this lack of solid evidence, many in the Church—perhaps a majority in its leadership—shared Galileo's view that it was very likely true.

The interesting question that arises out of this historical fact is why did the Church formally and vehemently reject Copernicanism, even though it harbored strong suspicions of its validity? To ask that question is to begin to realize that Galileo's dispute with the Church was not about Copernicanism per se. In other words, it was not about whether the Earth moves. What, then, was it about? The answer to that question is the subject matter of this book, but it can be stated here in a nutshell.

The dispute was over two conflicting views of the nature of truth and reality and about the roles religion and science ought to play in defining the world we live in. Of far more fundamental concern to the Church than the details of the Copernican hypothesis was Galileo's belief in the reality of number, his conviction that the universe was essentially a mathematical entity, in some literal way composed of numbers. The Church, bolstered by Plato, Aristotle, and nearly two thousand years of theological thought, denied this, on grounds that it excluded the possibility that there was an ultimate goal and purpose to existence. For the Church, a mathematical, mechanistic interpretation of nature could never be more than a model, an intellectual artifact. Between theory and reality there would always be a gap that could not be bridged by human reason.

Galileo and his opponents in the Church understood the true nature of their dispute very clearly and explicitly; it is the modern myth of Galileo that loses sight of its real significance. The argument about the nature of reality and what we can truly know nevertheless remains the principal bedevilment of modern civilization, for as Rousseau said, what we think we know, but do not, harms us far more than what we do not know. It is here, in this implacable difference of opinion, that we can identify in its most basic form what I have called Galileo's mistake.

In my exploration of the myth I discovered there are many Galileos: Galileo the dutiful eldest son who made great personal sacrifices to support his mother and siblings after his father's untimely death; Galileo the truant lover who left the mother of his three children when social status beckoned; Galileo the father of two daughters whom he shut away in a cloistered convent at an unconscionably early age; Galileo the man of culture who loved music, art, and literature, especially the classics, and who rejoiced in the delights of the cellar and the table; and Galileo the scientific and philosophical polemicist, who had great power as a writer in the Italian vernacular and loved to flex his literary muscles in the cut and thrust of debate.

There are at least two "scientific" Galileos as well. First, and least familiar to most people, is Galileo the physicist, whose work in the mathematical and experimental study of mechanics and dynamics earned him the well-deserved title of father of modern science. His masterwork in this area is titled *Discourses on Two New Sciences*, and it was published in 1638, five years after his famous trial by the Inquisition and four years before he died of a fever on January 8, 1642, in his villa in the hills overlooking Florence. At the core of this work was Galileo's novel approach to description and interpretation of natural phenomena through mathematics. In pursuit of this new discipline he fatally undermined the physics of Aristotle, which had long ago achieved the unassailable status of ordinary common sense. Furthermore, in merging mathematics and physics, he was able to see that the laws of physics familiar to us here on Earth also apply in the celestial realm. Previously, these had been thought of as distinct domains, governed by separate laws. In pursuing his mechanical studies he also developed the modern idea of the experiment (which he called the "ordeal"), constructing many laboratory devices, including various inclined planes and pendulums.

The centerpiece of his mechanical study was his investigation into the laws of motion. In his experiments with falling bodies he discovered that acceleration takes place continuously from the moment of release, and that all bodies fall at the same rate. His work paved the way for Newton's monumental theoretical structure of physical laws. Although Galileo published his findings in this field late in his life, many of his papers circulated privately in draft form, and he established a Europe-wide reputation as a leading mathematician and physicist early in his career.

The second "scientific" Galileo is the one that most often defines him in the popular mind: Galileo the great astronomer. Here, his reputation rests on less secure foundations. Galileo might almost be called an accidental astronomer. His main field of interest was physics, and though he lectured in astronomy at university, it was the ancient system of Ptolemy he taught, in which the Earth is at the center of the universe. His initial adoption of the telescope can be ascribed to his perennial need for money to support his family: he saw in it an instrument he might profitably manufacture. When he finally turned it to the night sky, his discoveries were many and spectacular, and they played an enormous part in the discrediting of the old Ptolemaic beliefs. But his endowments to astronomy are confined almost entirely to these observations. He made no lasting contribution to astronomical theory and was in the thrall of a stubborn conservatism that would not allow him, for example, to accept the idea that planetary orbits could be anything but perfectly circular. His contemporary Johannes Kepler far outshone him as a theorist, and there were others, including Jesuit astronomers, who were equally competent observers.

Nevertheless, it is as an astronomer that he is mainly remembered today, for it was his interpretation of the discoveries he made with his telescope that brought him into conflict with the Church and led to his infamous trial in 1633. I must emphasize immediately that it was the *interpretation* of these discoveries, rather than the discoveries themselves, that was the cause of all the trouble.

Although Galileo's brilliant successes in mechanics, or what we would call physics, were ferociously disputed by the entrenched Aristotelian academic establishment, they went largely unchallenged by the Church. This was not because the Church was uninterested or lacking in expertise. It was instead because in this area of study he was able to avoid metaphysical issues. He could stick to questions of "how" and ignore the "why." In astronomy, that was not possible because Heaven occupied astronomical space, and it was in the ordering of the cosmos that the mind of God could be discerned. Moreover, where Scripture touched on astronomy, it appeared to contradict his conclusions.

The journey of discovery that is reflected in this book took me to the places where Galileo lived and worked—to Venice and Pisa and Siena and Florence and finally to Rome. Each city was enlightening in its own way, but Rome was an epiphany. Rome, as I discovered, is a city where answers can profitably be sought to any and all of the questions troubling Western civilization. There are clues everywhere: in the relics of Roman imperialism and republicanism; in the scars of barbarian invasion and more recent European wars; in the remains of early Christian civilization; in its monuments to the Age of Faith and the Renaissance and the Age of Reason; in the vestiges of fascism and the dark side of scientific advance; in the monuments to the triumphs and failures of secular humanism and the enduring presence of the Christian Church. The evidence is all here in bricks and mortar, in travertine and marble, in books and paintings, in libraries and galleries and gardens and palaces and villas and churches that could occupy several lifetimes in exploring. Here, more than anywhere else, it is possible to look past centuries of prejudice and apologia to see the Church of Rome in its true historical perspective—magnificent, flawed; its leaders often brilliant and saintly, sometimes fatuous and contemptible; repository of a splendid vision that inspired the city's and our civilization's most wonderful art and architecture and built some of its blackest dungeons.

For many years during the sixteenth and seventeenth centuries, when the Medici family ruled the city-state of Florence, the Villa Medici with its storied gardens was the seat of its embassy to Rome. It was thus in this glorious Renaissance palace that Galileo, as chief mathematician and philosopher to the Medicis, lived through the most dramatic events of his career. His first sojourn came in 1611. He was then a middle-aged professor of mathematics at the University of Padua who had achieved a modicum of fame in academia for his work on mechanics and hydraulics. But he had recently begun looking into the night sky with a remarkable invention constructed of two lenses enclosed in a long tube, which had the astonishing effect of making distant objects viewed through it appear close at hand. He had just published a small, exciting book called *The Starry Messenger*, in which he described and interpreted the amazing discoveries he had made with his telescope. More stunning than the discoveries themselves were the conclusions he drew from them. At a stroke, he claimed, they crushed the prevailing wisdom that the Earth and heavens were distinctly separate realms of existence, one imperfect and subject to decay and the other perfect and immutable. Furthermore, they lent persuasive support to the recent and highly controversial hypothesis of Nicolaus Copernicus to the effect that the Earth, rather than being stationary at the center of the world, rotated on its axis and orbited the Sun. For several months Galileo was lionized in Rome by clergy and nobility alike with an enthusiasm that amazed even him, a man of no small ego.

Predictably, once the initial amazement at the idea of mountains on the Moon and satellites orbiting Jupiter and the sheer, unimaginable number of stars in the Milky Way had subsided, reaction set in. His next visit to Rome in the winter of 1615–16 was devoted to defending his own name as a good Catholic and to derailing attempts by Church conservatives to ban the writings of Copernicus. In 1624 the long season of conservatism within the Vatican appeared to have ended, with the election of a cultured and sophisticated new pope, Urban VIII. Galileo visited Rome to pay his respects and promote, as vigorously as he dared, both Copernicanism and his own revolutionary theories of science. Six years later, he spent two months in the city arranging for publication of his most famous work, *Dialogue on the Two Chief World Systems: Ptolemaic and Copernican*. Urban VIII again welcomed him warmly.

In 1633 Galileo paid his final visit to Rome, under dramatically different circumstances. He had been summoned by an angry pontiff to stand trial before the cardinals of the Inquisition for "vehement suspicion of heresy." Exactly what was behind this drastic change in his fortunes has been the cause of centuries of historical debate. I hoped that my own sojourn in the city would help me to solve the puzzle—if not for all posterity, at least to my own satisfaction.

TWO

Pope Paul V • A Time of Crisis
Doctrinal Revolutions

The Borghese family, who loom large in the story of Galileo, has its roots in the city of Siena in Tuscany where, from the twelfth century onward, its members were well known as jurists and ambassadors. They migrated to Rome in the sixteenth century, when the Medici family of Florence forcibly annexed Siena to their domain. In Rome, the Borgheses continued to flourish, and in 1605 the family secured both fame and fortune when the studious Camilo Borghese was elected pope, taking the name Paul V.

Paul V occupied the papacy during much of Galileo's life, up to and including his rise to fame as an astronomer. It was he who initially challenged Galileo's Copernican sympathies, and who ordered Copernicus' great book, *De Revolutionibus*, withdrawn for "corrections." He is portrayed in histories of the period as a foul-tempered martinet with little sympathy for new ideas in art, religion, or anything else. The Tuscan ambassador to the Holy See described him as "a man so averse to anything intellectual that everyone has to play dense and ignorant to be in his favor."[1] But that was after the pope had embarrassed Tuscany, along with Savoy, Genoa, and Naples, forcing them to back down in a power struggle over the secular limits of Church authority. Giorgio de Santillana, author of an exciting if biased work on the trial of Galileo, says of Paul V, "[he] was not an open mind, nor much of any kind of mind. He was a strong and somber executive, a canonist by training, by temperament doctrinaire and inflexible. As he once said, he preferred new jobs for workmen to new ideas from scholars."[2] Certainly, he had little interest in the sciences. At the same time, it was Paul V who, early in his papacy, gave St. Peter's Basilica its final structural shape through architect Carlo Maderno, making the great church a focus of artistic patronage for generations to come. He began work on the famous Borghese Palace, enlarged the Quirinal and Vatican, extended the Vatican Library, and began a collection of Greek and Roman antiquities. He restored the aqueduct of Augustus and Trajan and made it the source for many fine new fountains in Rome, including one for the Jews in the Piazza de la Sinagoga. He supplied encouragement and resources to missionaries in Asia and the New World and granted Chinese Christians the right to say Mass in their own language. He was, in short, not quite the one-dimensional man the offhand aspersions of popular historians or hostile diplomats often suggest.

The throne of St. Peter was not a comfortable one in the early seventeenth century. There was dangerous unrest in Germany between Catholics and Protestants, which would soon ignite the Thirty Years War. There was trouble with King James I of England, who had required Catholics to take an oath of allegiance to the British crown, placing secular above ecclesiastical authority. There was the troubling emergence of the European nation-state, a phenomenon encouraged by Luther's Protestant revolution and the resulting disengagement of the territorial nation from religious and moral oversight, which evolved into the doctrine of the separation of church and state. Christendom was undergoing a process of fragmentation, and the new nation-states saw themselves as separate moral entities with their own, parochial missions. It was a time when Catholic Spain and France presented more or less continual crises with their mutual penchant for warfare and territorial expansion at the expense of their neighbors, and the Holy Roman Empire* posed an ever-dangerous challenge to the notion of supreme papal authority over secular rulers.

---

*The Holy Roman Empire (the name is a convention of modern historians) was present in Europe from the eighth century until 1806. The Roman Imperial title

Within the Catholic Church itself there was the need to maintain the momentum of the highly successful program of reform and militancy that grew out of the Protestant schism. In 1545 Pope Paul III,† to whom Copernicus dedicated his epoch-making masterwork *De Revolutionibus*, had called into session a Church council in the Italian city of Trent. It was to continue its deliberations until the year before Galileo's birth. The authority of such councils exceeded even that of the pope. This one had been charged with undertaking a complete reexamination of Church doctrine and practice in response to the disastrous split with Calvin and Luther known as the Reformation. The goal was to restore clerical discipline and give the faithful a renewed sense of security by providing a highly structured and clearly interpreted theology. There was no shortage of abuses in need of correction: cardinals, bishops, and priests incessantly absent from their domains; widespread concubinage, drinking, and hunting among the priesthood; ownership of private land by priests; rampant granting of dispensations in return for money and privilege; shocking levels of illiteracy and incompetence among junior clergy. While much was accomplished, the so-called Counter-Reformation was not without its costs. The Church was seized by "an atmosphere of obsessional control over detail, endless doctrinal clarifications by councils, synods, and theologians, suspicion of deviancy, and a proclivity for inflexible, legalistic remedies in areas of social conflict. . . ."[3]

---

had lapsed in the West after the fifth-century barbarian invasions, to be revived by Pope Leo III in 800, and conferred on Charlemagne. When the Carolingian line died out, the crown was annexed by German kings whose dynasties ruled from the tenth century onward. Originally allied with the papacy, the empire engaged in a long series of debilitating struggles with Rome over secular versus religious authority. It was further weakened by the Protestant Reformation and the resulting split between Catholic emperors and Protestant princes. Beginning in the fifteenth century, the throne was occupied almost continuously by Habsburgs. The last emperor to be crowned by a pope was Charles V, crowned by Paul III. The German empire of 1871–1918 was called the Second Reich (empire) to denote its descent from the medieval empire; Hitler's Third Reich made a similar claim.

† It was Paul III who managed to persuade Michelangelo to return to the Sistine Chapel to paint *The Last Judgment* on the altar wall. At its unveiling the pope is said to have fallen to his knees, overcome by emotion.

To add to the confusion and ferment of the times, the entire balance of economic power within Europe was shifting like the cargo on a listing ship, thanks to the opening up of the New World. The forging of sea links to the Americas led to a migration in the focus of European trade from the Mediterranean to the Atlantic seaboard, and countries without Atlantic ports suffered competitively. More than one hundred thousand Europeans, mostly English and French, were now living in North America. London became the fastest growing city in Europe. Italy, once as much a leader in trade and manufacturing as in culture, and one of Europe's leading exporters of manufactured goods, was headed into economic eclipse that was to last for two centuries.

It was in this period of waning fortunes and political influence in Italy that Paul V had been crowned, in 1605. The young Galileo was then a junior lecturer at the University of Padua, teaching Ptolemaic astronomy and mathematics and privately tutoring the sons of aristocrats. Prince Cosimo de' Medici of Florence was, in that year, one of his prize students. At the same time, Galileo was quietly studying Copernicus and honing the ideas that would challenge and ultimately topple the age-old authority of Aristotle. It was a cautious and conservative Paul V who was thus to be confronted with the first tremors of the Scientific Revolution.

In the long history of the Church, this latest revolution in thought was only the most recent in a series of philosophical crises. The first of these arose from Greek philosophy, which had both challenged and shaped Christian thought from the earliest days of the Church. In the freshness of its youth, Christianity had been forced to come to terms with Plato, in the form of the Neoplatonism of Plotinus, a philosopher widely influential in third-century Rome.* The philosophical stew called Neoplatonism was, in its scope and interests, effectively a religion. The great theologian and philosopher St. Augustine of Hippo, recognizing this and seeing the competitive threat it posed to Christianity, labored mightily in his writing and teaching to enfold Plato within Christian thought. In success-

---

*Neoplatonic thought in turn was tinged by the early number-mysticism of Pythagoras (c. 582–c. 500 B.C.) and influenced by Stoicism, a quasi-religion that placed heavy emphasis on personal ethics. Stoicism was popular among Roman leaders of the first and second century, notably Seneca and Marcus Aurelius.

fully doing so he neutralized the impact of Neoplatonism and at the same time transformed Christianity from a collection of beliefs and instructional narratives into a true systematic philosophy, one that would become an increasingly subtle and powerful instrument of knowledge and understanding over the succeeding centuries.

Eight hundred years later the Church would undergo a second, similarly wrenching, philosophical encounter, this time with Plato's student, Aristotle, when that philosopher's works were discovered anew by Europeans. But the engagement was different in its essence, since, while Plato had represented a competing religious view, Aristotelianism was a philosophy—almost, we might say, a science. It relied principally on observation and classification (as opposed to contemplation and revelation) in gaining knowledge. Nevertheless, the task facing the Church was once again to incorporate an ancient, coherent, and undeniably useful and revelatory body of thought into Christianity. A champion was found in St. Thomas Aquinas (1225–74). He was strikingly successful, one of those prodigious figures whose talents and abilities seem to match perfectly the requirements of the historical challenge facing them. He dragged himself away from Paris after five years of labor, exhausted and near death, but having through monumental intellectual and literary effort tamed Aristotelianism.

His achievement would result in relative doctrinal harmony throughout Western Christendom for more than three hundred years. Even the Protestant Reformation would not shake the essential authority of Thomist Aristotelianism, which was accepted by Catholic and Protestant alike.

And then, in the early seventeenth century, Galileo and the novel blend of old and new ideas he represented brought upon the Church its third and most profoundly dangerous philosophical crisis. This time, though many tried, no philosophical prodigy was able to save the day by reconciling the old with the new. The world, it seemed, was determined to become modern, at any cost.

THREE

A Dialogue • Science's Motives
On Curiosity • A Negotiation

While in Rome, I had the good fortune to bump into my friend and one-time student Berkowitz, to whom I had given Hans Blumenberg's *The Legitimacy of the Modern Age.* I was curious to know what he thought of the book, so I asked him about it over dinner one evening.

"Frankly, I haven't had much time for reading. I brought it with me, but I've only had a chance to glance through it. Pretty dense-looking stuff," Berkowitz said.

"Every paragraph contains a pearl," I insisted. "He was an amazing guy. You get the impression he knows everything there is to know about Western civilization. I heard that he lived in a castle somewhere in Germany and devoted his life to scholarship. I guess that's what it would take—not having to worry about the daily routines of kids and cooking and bill-paying and all that."

"Bliss," Berkowitz sighed.

"I think it's one of the great books of the century. It seems to me he's put his finger on the main problem."

"Which is?"

"You want me to summarize a six-hundred-page book in one sentence?"

"Why not?"

"Okay, then," I said, "how's this? He'd say that the problem is that we've lost track of the proper motive for doing science."

"Which is?"

"Which is the promotion of human happiness."

Berkowitz paused to consider this. "Which species of happiness did you have in mind?"

"I mean the old-fashioned kind. Happiness in the sense that you have everything that's worth having in life. Not everything you want, necessarily—that's hedonistic happiness—but everything that's worth having."

"Worth having according to whom?" he asked.

"According to some objective standard. I'm sure we could agree on a few main items—knowledge, aesthetic pleasure, friends, and so on—but that's getting way off topic."

"But you're implying that modern science has some other motive for what it does. What other motive would there be?"

"We *think* that the promotion of happiness is what science aims for, but as the man says, the reality of the world we live in is veiled by misleading ideas. You only have to read a newspaper to see that science has its own agenda. When it does serve humanity's real interests, it does so more or less by accident. I mean, scientists and their apologists have this presumption of service built into all their rhetoric, but if you examine it, it's meaningless. They tell us, 'Science is performed for the welfare of the human species.' But when you question a particular development—for example, some particularly gruesome 'breakthrough' in biotechnology—when you ask, 'Can this really be good for us?' they shut you up with 'Of course it's good for us. It has to be because it's science, and what science does is good for us.'"

"Okay," said Berkowitz, "let's take science to mean knowledge, which is what it actually does mean, right? Naturally there's a presumption that the pursuit of knowledge is always a good thing. You know what they say, 'The truth shall make you free.'"

I had to laugh. "You're quoting Jesus Christ there! I suspect he had in mind something other than scientific fact when he used the word *truth!*"

Miffed at my laughter, Berkowitz waved it away with the back of his hand. "Whatever," he said.

"Anyway, the thing is that what science calls knowledge is often misinformation. It is often incomplete, or just wrong. And it most certainly is never 'the truth' about anything other than itself."

"What. . . ?"

"Let's leave that for now. The point I'm trying to make is that science has simply defined itself as 'good,' and we've bought the hoax for the past 350 years."

"I hate to be the one to tell you this," Berkowitz said with a smug smile, "but we'd all be in deep shit without science. A lot of us wouldn't even be alive today."

"I'll grant you that we can't get along without science. But that, as Blumenberg says, is an effect produced by science. Science has made itself indispensable. Quite clearly, though, it wasn't always that way. Science and the technology that arises out of it—the kind of science and technology we're accustomed to—is an invention less than four hundred years old. Before that, people got along without it. The ancient Greeks clearly thought they had a pretty comfortable life—in fact, they thought that was a prerequisite for doing the kind of philosophy they did. Or take the Renaissance and all the wonderful art we've come all this way to see, and all the sixteenth-century humanist philosophy and literature we read—it all happened before the Scientific Revolution.

"Until the Scientific Revolution, nobody thought that the purpose of science was to make life possible—they believed it was to make life happy. Aristotle took it for granted that life was more pleasant for the person who knew things than the one who knew nothing. Socrates before him said that the unexamined life isn't worth living."

"Well, I still don't get it," Berkowitz said. "What is their motive, then?"

"Whose motive?"

"Scientists'!"

"Oh yes. I would say, in a word, self-indulgence."

Berkowitz rolled his eyes and leaned far back in his chair. "That's ridiculous. Scientists, the good ones, are selfless people. Look at Einstein. Look at Madame Curie. Look at Faraday!"

"Exemplary people," I agreed. "But that's not the point. If you go back in time before the Scientific Revolution, before Galileo and Descartes and Newton, you find there was a different attitude toward curiosity about nature. This goes back to St. Augustine and beyond him to Socrates and Plato. Augustine called it 'pious inquiry.' The idea was that there needs to be a link between curiosity—curiosity as the pursuit of theoretical knowledge—and morality. Science is dangerous without moral restraints."

Berkowitz had been listening a little distractedly, his legs stretched out under the table, twirling the stem of his wineglass back and forth between his thumb and his forefinger.

"You're going to have trouble convincing me that curiosity is a bad thing," he said. "Next you'll be telling me the Inquisition was perfectly justified in torturing poor old Galileo!"

"They didn't torture him."

"Of course they did."

"They didn't torture him. He was questioned under formal threat of torture, because that was the routine. It was supposed to ensure that you told the truth. The idea that he was actually tortured probably comes from a nineteenth-century writer named David Brewster, who said so in a book called *The Martyrs of Science*. Something of a bigot, if you want to know the truth. But the story is certainly false. Nor was he shown the instruments of torture, as some more recent accounts have said. It was just a formality."

"Okay, if you say so, but I still can't buy the idea that curiosity is a bad thing."

"Not *always* a bad thing, or even most of the time. Think of it like lust—something that needs to be kept in check but definitely has its season."

He smiled and sat up. "Okay, you've got my attention."

"Right up to the seventeenth century, as I say, there was a notion abroad in the world that curiosity and inquiry are not always and unconditionally to be admired and encouraged. It was fine to inquire into the areas of nature that were accessible to us, and these were defined as the areas from which we could gain knowledge that would make human life physically better, more secure and comfortable. In other words, it is okay for science to look at things in the world from the point of view of the usefulness inherent in them. But curiosity for its own sake—the pursuit of worldly knowledge for its own sake—was frowned upon. And there were degrees of disapprobation. At best, it was a waste of time, because nature was ultimately unknowable. At worst, it was a damaging self-indulgence that led people away from the true roots of their existence and the true goals of wisdom, which were not scientific at all but religious and philosophical. Science was an amusing and often useful side-show. But *only* a sideshow, and the danger of distraction was serious. Wisdom was to be gained by looking inward, not outward. It's an idea that goes all the way back to Socrates. Christians—and nominally at least, that included most of the population of medieval Europe—believed that after Christ and his teachings, everything that was really important to know, was known."

Berkowitz snorted in what I took to be derision. He seemed suddenly out of patience. I hurried along, determined to complete the point.

"Look," I said, "it's one of those ideas that's difficult to understand until you know where it comes from. It's basically a pre-Christian, Neoplatonist point of view, and it has to do with their conception of the soul. They thought that there was an overall world soul. Each human soul originates in the universal soul, and as long as it dwells in the human body it's at risk of being lost. But it can protect itself by thinking back to its origins, by keeping in touch with the mother ship, as it were. If it fails to do that, it will become dissatisfied and restless, and this manifests itself as curiosity. Curiosity is not so much interest in the object but a symptom of the unrest and dissatisfaction that comes with being alienated from the world soul. You might say it's a sign of a loss of focus. The Neoplatonists said that danger to the soul came from seeing the 'many'—in other words, the things of this world—rather than the 'one,' which was the universal soul. Medieval Christianity picked up this basic idea through St. Augustine."

Berkowitz was examining his fingernails and shaking his head. I pushed on.

"It really is a very difficult concept for the modern mind to grasp—that not all knowledge is good knowledge, that curiosity can be counterproductive."

"No kidding," Berkowitz chuckled. "For a start, how can you possibly know where to set the limits on the pursuit of knowledge? Obviously, the trouble with setting limits is that you never know what tasty morsel is waiting around the corner, do you? I mean, how is it possible to know for sure what line of inquiry is going to turn out to be useful and what isn't? I mean, Alan Turing was looking for a way to confirm Gödel's bloody theorem when he came up with the rules for building digital computers, was he not? Face it, half the important discoveries in biology and chemistry were made by people just tinkering around in their labs, or recognizing a key to some breakthrough in some unrelated mass of data. The fact is, there is no way to know where limits to inquiry ought to be drawn. That's why it's best not to have any limits at all."

He had me backed into a corner, and I was getting irritated.

"You might be able to sell me on the first part of that statement," I told him, "but not the second. There are plenty of areas of life where we don't have the kind of definitive knowledge we'd like to have before making rules. But we go ahead and make rules anyway, because rules are needed and perfect knowledge is a mirage. That's practically the whole history of jurisprudence. And then if the rules have to be revised in light of new information, they can be changed. Errors will be made, progress will be slowed from time to time, but the up side is that change can be managed for the benefit of people, rather than people continually having to adapt to change that comes at them out of left field."

Berkowitz did not reply but made a poor attempt at concealing a yawn with the back of his hand. "It's been a strenuous day," he said.

"Obviously I'm not engaging you here," I said to Berkowitz. "You don't seem to be getting the point. What I'm saying is that our attitude to science has changed in recent history, and probably not entirely for the better. We used to think that knowledge made for happiness, as Aristotle said. That's why we pursued philosophy. We loved knowledge because it made us happy by reconciling us to our place in the universe. Then scientific knowledge got separated from the rest of philosophy. That happened with Galileo and Bacon* and the rest of that crew. For them, science became a communal enterprise that was only incidentally or secondarily, if at all, interested in individual fulfillment and happiness. The only knowledge humanity needed for happiness, as far as they were concerned, was what

---

*Francis Bacon (1521–1626), an English contemporary of Galileo's and a champion of empiricism, induction, and experiment—the scientific method.

would give us the upper hand over our physical surroundings. So the knowledge science sought was the knowledge necessary to subdue nature. As Blumenberg says, for the new scientists, the recovery of paradise was not supposed to yield a transparent and familiar reality anymore, but only a tamed and obedient one.

"According to this scientific scheme of things, a person no longer needs to understand himself in relation to reality. All that's required is that everybody's joint contributions to science add up to a state of stable domination over this reality. That way, everyone can benefit, even though no person completely understands the total picture. That way, individuals achieve a kind of happiness with only limited knowledge of reality. And the fact is, although we know a great deal more about the world than we ever did before, 'we' is a collective pronoun. It's not 'we' as individuals who have all this knowledge. It exists only in our institutions, like universities and libraries and databases. And it's employed and administered not by ordinary people like you and me but by initiates and specialists.

"When you think about it, part of what went on after the Scientific Revolution was a redefining of happiness to suit science's more limited goals and aspirations. Not to mention its more limited view of what constitutes reality. For science, happiness meant physical security and success in making life possible. Before that, it had been something more, some innate reward in the grasping of truth—a reconnection with the world soul, you might say. Something transcendent.

"What you get as a result of the dividing of the ways, that fork in the road that came with Galileo and Bacon, is the evolution of a science that is no longer the disinterested or impartial pursuit of knowledge for the sake of human happiness but an industry like every other industry. I think that if you were able to ask them, you'd find that the early champions of science in the seventeenth century would be astonished and not a little disturbed at what they made. Galileo and Descartes, I think, would be aghast. Because nowadays if you want to understand modern science, you have to look to the goals and aspirations of industry, not philosophy. That is why it is legitimate to want to put some controls on it. If it is not too late already," I concluded.

Berkowitz and I said a somewhat testy good-night with a

promise to meet again over lunch or breakfast in a day or two.

On the way home I was thinking, How *do* you make rules for science?

I saw him the next morning. Whether by accident or design I could not tell, but Berkowitz was outside my hotel at a small open-air bar and restaurant at the lip of the escarpment across the road from the Villa Medici. His features were more Greek than Semitic. With his widow's peak, heavy beard, and faintly bird-like features, he resembled the Hollywood actor Nicolas Cage, though without all the muscle. He had once explained to me that his name had been proudly adopted by his grandfather, a wartime foundling of unknown provenance who'd benefited from the compassion and generosity of a family of Macedonian Jews. Berkowitz described himself as "an evangelical atheist."

The man I saw before me was a far cry from the person I'd met years earlier. As a middle-aged victim of the 1990s mania for corporate anorexia, he'd been downsized out of a long-time job and after a series of dispiriting interviews with an "outplacement consultant," left unemployed and confused by his status as a statistical "human resource." He showed up in a university extension course I taught as part of his retraining—"reprogramming," he called it.

"What the hell is a human resource?" he'd wanted to know. "I always thought I was a human being!" He'd struggled with a profound depression.

The experience left him with a permanent short fuse and an abiding cynicism for the kinds of social graces required of corporate functionaries. He could be insufferably rude. For all of that, though, it had always been clear he was anything but a loser. To no one's surprise, his fortunes had taken a decided turn for the better in recent years. He'd landed a senior public service job, and soon after that had quit to start a spectacularly successful business in the health care field. I had heard he'd come into an inheritance as well. He no longer had any real need to work. As for our friendship, the two of us had somehow never managed to transcend our initial student-teacher relationship despite the fact that his income dwarfed mine, and he was a man of no small erudition when he cared to show it.

"So this book of yours," he began, disdaining the formalities.

"Am I in it?"

"I would like you to be. With your permission, of course."

He looked out over the city, saying nothing. I continued, "I intend to make a lot of use of dialogue to help the philosophical bits along."

"Like Plato," he suggested.

"Well, yes, but more to the point, as Galileo did in his books. He invented characters to represent various opinions and gave them the names of friends he'd admired in real life."

"Then you'd better fill me in on what it's about so I can do my homework. I mean, I know it's about Galileo and the Church, but what's your ulterior motive? You're not just telling a story; otherwise, you wouldn't need me."

I did my best to conceal my delight. Not many characters, in my experience, were as good at adding the tension and zing of lively dialogue to a work of nonfiction. But Berkowitz could also be difficult to deal with if he felt he had the upper hand in the mysterious transaction between the writer and his creation. I had no intention of letting him know just how important was the role I had planned for him.

I told him that I did indeed intend to retell the story of how Galileo came to be summoned before the Inquisition and charged with "vehement suspicion of heresy," because I felt that the events of the case had been misrepresented in so many accounts that it was time they were set down afresh, just as they happened. But the chronology, I said, would be just one of several different threads running through the book. And behind it all, behind the history and philosophy and science, was, indeed, an ulterior motive.

"Which is?"

"If you learn anything studying the Galileo affair," I said, "it's that the same body of factual information can be interpreted in radically different ways by different commentators. The difference is in the preconceptions and prejudices each of them brings to the subject matter. My own take on the story is backed up by current thinking in experimental physics and the philosophy of science. When I look at the historical record from that perspective, certain things just seem to fall into place, like bits of a puzzle. The record is very complete, and it's pretty straightforward if you just take it at face value."

"And I suppose you're going to write the perfect objective account."

"There's no such thing. I've just told you I have my own biases. But I intend to be up front about them."

"And others weren't?"

"Some were. For instance, the Catholic clerics and laymen who've written commentaries are obviously approaching the subject from a point of view that is at least somewhat sympathetic to the Church's position. That's not to say they distort the facts or anything like that, but they will have a predictable approach to interpreting what those facts mean. And on the other hand, you have some decidedly anti-clerical, anti-Catholic writers who've been very hard on the Church and more or less deified Galileo. Some of these writers have been anything but scrupulous with the facts. This is where people get the idea that Galileo was tortured and thrown in prison and so on, and that the Church was run by a bunch of ignorant and venal prelates who wanted to suppress the truth in order to maintain their political power."

Berkowitz smiled. "That's pretty much the version I grew up with."

"Me too. Those obvious kinds of biases are fairly easy to identify. But there's another, more insidious, prejudice that almost always goes unnoticed, and that's the one I'm really interested in avoiding myself. Actually, it's two closely related biases, and they're pretty much universal.

"The first is an automatic reflex assumption that science and reason always trump religion and revelation. That real truth comes, as it were, out of a test tube and not out of some holy book or other —or some mystical experience. Even Catholic writers will often accept that this is the case when it comes to knowledge of the natural world."

"And you think this is wrong, incorrect?"

"I think that it is wrong to grant science exclusive right to say what is true and what is not true about the world, yes."

"I see. So you're going to be an apologist for the Church, then. You're going to join the great tradition of Catholic casuists." He said it with a smirk, and I knew it was intended to annoy me. I refused to rise to the bait.

"I think you know me better than that," I responded. "Though I'll grant you my argument may well seem like an apologia from time to time, simply because the view we currently hold of the world is so blindly off-kilter, so unquestioningly pro-science. Common knowledge is that religion can tell us nothing useful about the natural world and should keep its nose out of science. I'll be arguing that there is a legitimate place for religious insight in the pursuit of science—in fact, I'll be arguing that science is actually a kind of religion itself, but that's another story. But if you're suggesting I'm going to be an apologist for anybody, you're wrong. And as a matter of fact, I resent the suggestion."

"What's the second bias?" he asked.

"The second bias is an unexamined assumption that human history is a matter of linear progress from primitivism to sophistication in all things. According to this point of view, there may be periods of backsliding, like the so-called Dark Ages, or stagnation, like the Middle Ages, but by and large, the idea is that we're moving ahead in a straight line.

"It's a relatively new idea, not more than about three or four hundred years old. Prior to the seventeenth century and the Scientific Revolution that Galileo helped to launch, most people thought of history as cyclical rather than linear. That meant that it was possible, even likely, that there had been periods in times past when men and women were in one way or another superior to ourselves and that there was much to be learned from them. The cyclical view also implies that there are limits to human potential, in the sense that we will never build heaven on Earth. We are in a circular system. We can certainly improve things, but they will never be perfect. At the same time, though, there are no limits to our potential in terms of spiritual development. We can join the infinite. That's why, from its earliest beginnings in Greece, Western philosophy has had a fairly continuous consensus around the idea that the way to live was to go for spiritual development, where the potential was real, and steer clear of too much involvement with the material, because its benefits are fickle and its knowledge is illusory.

"With the Scientific Revolution and the Age of Reason, knowledge and enlightenment came to be associated with the material rather than the spiritual, in other words, with the steady accumulation of factual information about nature. The idea was that eventually we should be able to accumulate enough facts to construct a coherent description of everything. We would know what God knows. This is the Enlightenment project—the project of the Age of Reason—and we've persevered with it. It replaced a circular, finite, and inward-directed process with one that's linear, outward-looking, and essentially endless in scope, because the supply of new facts is inexhaustible."

"You said you had an ulterior motive," Berkowitz reminded me.

"And so I do. I want to make the point that badly needs to be understood these days, and that is that science is not the only legitimate font of knowledge and that it can and should be challenged on some of its most fundamental preconceptions. I want to lift its skirts and expose the rot in its foundations."

"So how does Galileo fit into all this?"

"I think that the story of Galileo is one of the foundational mythologies we've used to promote the Enlightenment viewpoint we've inherited. We have made saints and heroes of leading scientific, rationalist figures such as Galileo, Newton, Maxwell, Darwin, Einstein, Oppenheimer . . ."

"Bertrand Russell . . ." Berkowitz contributed.

"Russell, possibly, and scores of others. But Galileo is the big one. On the other hand, we have mythologized religion in general and the Christian Church in particular—since it is the important religious institution here in the West—as backward, anti-intellectual, and violently opposed to progress."

"But that pretty much was the case, wasn't it?"

"Only if you look at history through scientistic, rationalist glasses. Put on a pair of religious-spiritual glasses, and you get a different picture."

"You mean *scientific*."

"What?"

"You mean *scientific*, not *scientistic*."

"No, I mean scientistic, in the philosophical sense of—"

"Of science having a corner on 'real' knowledge. I remember now.* But are you trying to tell me that different spectacles change

---

* *Scientistic* is a term coined by philosophers of science to designate a mode of

the reality they help you to see?"

"It's not just me saying that. It's the scientific establishment. The leading-edge experimental physicists since the beginning of the twentieth century pretty well all agree. Not to mention the principal philosophers of science for at least a century before that. All scientific knowledge is culturally conditioned. None of its laws or facts are, strictly speaking, objective. I find it ironic that at about the same time the Church decided to rehabilitate Galileo in the nineteenth century, European philosophy was busily demolishing the philosophy behind his whole edifice of scientific empiricism."

"Empiricism?"

I had to smile, and I gave him the movie-maker's sign for *cut.* "Okay, that's great. But you're getting me too far ahead of my story. Can we leave it there, for the moment?"

He shrugged. "Fine. You now know where to reach me."

---

thought that assigns to science exclusive and unchallenged authority over truth. Scientism, to use the noun, is a mechanistic world view that comes down to us in line of succession and with increasing assertiveness from the great medieval theologians and philosophers Thomas Aquinas and William of Ockham through to René Descartes. Descartes, the seventeenth-century French thinker whom many call the founder of modern philosophy, saw the universe as a clockwork mechanism of cause and effect in which all living entities, except for humans, were mere machines. Humans were distinguished in this world of automatons by entertaining a ghost within the machine—the soul or spirit, which was their exclusive link to God. The worldview of modern scientism, according to which what cannot be weighed or measured does not exist, is essentially Cartesian. However, scientism goes one important step further than Descartes, in asserting that human behavior of all kinds can be attributed exclusively to biochemistry, and that the spirit has no reality except as a figment of our biochemically stimulated imaginings. In a scientistic world, to argue that science and religion are incompatible is to seriously understate the case. Religion, moral philosophy, aesthetics—all thought that concerns itself with things other than the material—is simply non-sense. Nonsense.

FOUR

Galileo the Aquarian • Ptolemy's World
Comets and Supernova • Kepler's Genius
The Uses of Hypotheses

Galileo Galilei was born into an aristocratic old Florentine family of declining fortunes in Pisa on February 15, 1564, the first of six (possibly seven) children. It was the year of Michelangelo's death and Shakespeare's birth. His mother was Giulia Ammannati of Pesscia. His father, Vincenzo, was a cloth merchant by trade and by passionate avocation a musician, composer, and musical theorist with a penchant for mathematics. He wrote several tracts on music in which he betrays some of the traits that were to distinguish his son's character, including a distrust of intellectual authority that often rose to contempt and a pugnacious temper. He is one of a group of musicians who regularly gathered at the Bardi Palace in Florence to plot the reform of contemporary music. The objective was to revive Classical Greek approaches to performance by finding ways to integrate music with drama so that the two would reinforce each other to produce a theatrical experience of unprecedented emotional power. The outcome of their activities was the birth of opera, which had its debut in 1597 in the great hall of the Uffizi Palace.*

*At about the same time, the ballet was being born in the royal courts of France.

Vincenzo's son Galileo was born under the sign of Aquarius, the Water Bearer. This fact is of no interest to modern biographers of the great scientist, but in an age when planetary influences on earthly life were taken for granted, it would certainly have been portentous for his parents and for the child himself as he learned about the world around him. They would have known that as an Aquarian he would be predisposed to an extraordinary open-mindedness and breadth of vision, along with an aloofness from the strictures of tradition and authority. The Aquarian, it was understood, was a seeker after truth; patient, dispassionate, untiring, the very model of the philosopher, frequently touched by brilliance. Aquarians were known to be affectionate by nature and could be expected to go to great lengths to comfort and care for those around them. However, the same planetary influences made them subject to the risk of being overwhelmed by their many interests, unable to manage the practical details of life, missing opportunities, and frittering away time and energy on minor details. Aquarians were inclined to be tactless, and their peace-at-any-price bias could easily spill over into moral and physical cowardice. They were apt to be flustered and bewildered when they blundered into conflict, striking out clumsily and blindly, motivated mainly by a desire to get the battle over as quickly as possible. As with other areas of authority and tradition, Aquarians were likely to be unorthodox in their relationships with the opposite sex, though they made exemplary spouses and parents if they could be lured into marriage—something that normally happened only late in life. In religion they tended toward a thoughtful and respectful skepticism, inclined as they were to a reluctance to accept authority and bemused by a deep understanding of the difficulty of finding certain knowledge of their own accord.

The world of nature into which Galileo was born differed only in minor points of emphasis from the medieval cosmos lovingly described two and a half centuries earlier by the great Italian Renaissance poet Dante Alighieri in his *Divine Comedy*. Although Copernicus had published his revolutionary theory thirty years earlier, it had made little impact and it was an essentially Aristotelian world that greeted Galileo, last updated by the astronomer Ptolemy in the second century A.D. The Earth was made of rocks, water, and air, which have weight, emit no light, and are constantly changing.

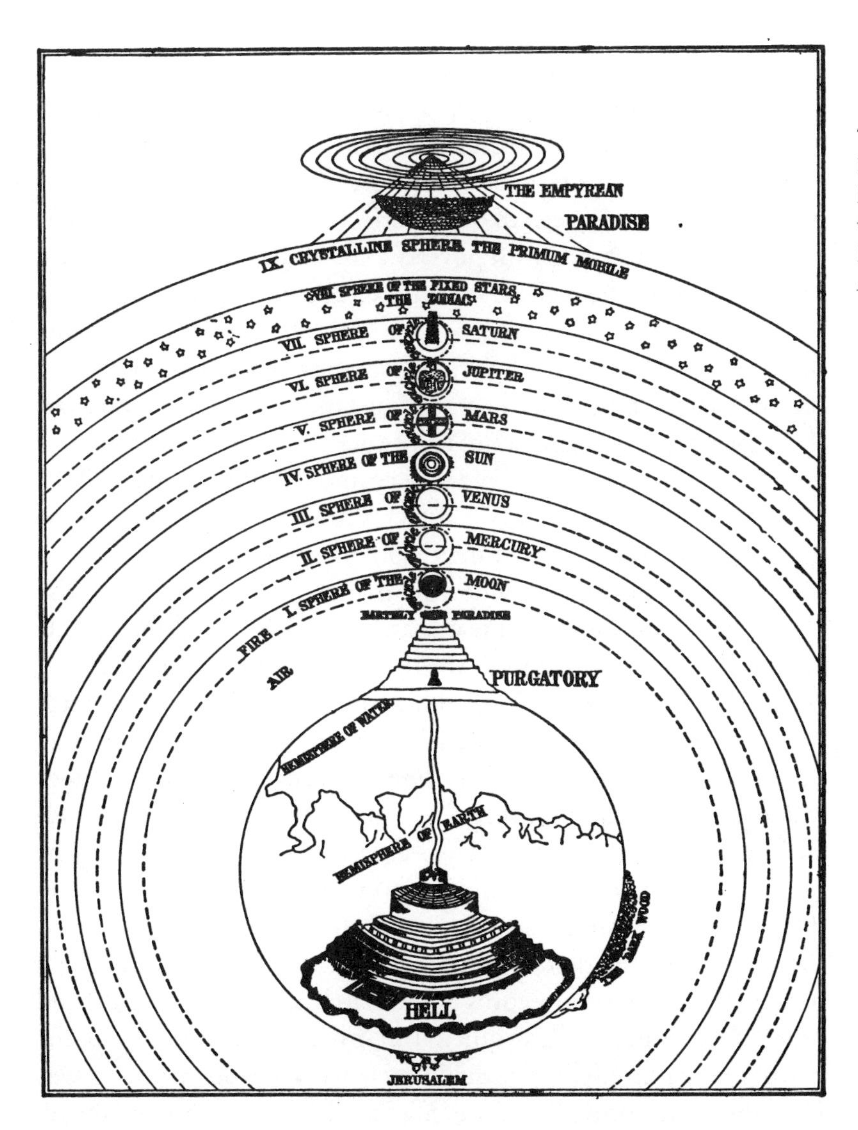

The Ptolemaic, Earth-centered world of poet Dante Alighieri.

The heavenly bodies, beginning with the Moon and including the planets and the stars, were made of an altogether different substance and were not subject to change. The planets, Dante said, were like jewels, resplendent in the reflected light of the Sun. Other writers differed on this detail, expressing the opinion that a planet was rather like a knot in a piece of wood, a thickening in the weightless, changeless, luminescent material that the heavens were made of.

The crystalline spheres upon which the planets rode, said Dante, were moved by intelligences, or spirits, or vessels of consciousness of various grades, a necessity created by Aristotle's common-sense conviction that the natural state of matter was to be at rest, and that to set it in motion and to keep it moving required effort. Among Dante's celestial spirits, the lowliest are the angels that move the Moon, which can be seen to be imperfect in its pockmarked visage, and which is associated with poverty and servitude. The Sun was at the other end of the scale. Long after Creation, Dante wrote, these spirits continue to influence life on Earth, their powers emanating directly (literally, as rays of energy) from the planets that they are associated with. Thus, Venus helps to shape the fate of lovers, and Mars that of warriors. As a Christian, Dante needed to find a place for free will in his deterministic astrological cosmos. In this he adopted the view that the human soul is the direct, unmediated creation of God and while the planetary influences may establish general dispositions in life, the soul provides the means for individuals to rise above this conditioning. This was an idea tolerated, if not warmly accepted, by the Church until the time of St. Thomas Aquinas.

We know almost nothing about the life of the great second-century geographer, mathematician, astrologer, and astronomer Ptolemy of Alexandria. But his *Almagest*, a comprehensive geological and astronomical synthesis, is among the most influential scientific works ever written. The *Almagest* was recognized as the pinnacle of European scientific achievement from the time of its rediscovery in the late Middle Ages well into the Renaissance. In it, Ptolemy gives a full description of the observable universe, in which he relies heavily on the records of Hipparchus (c. 197–120 B.C.) and other astronomers of antiquity. But his real interest lay in the mathematical modeling of the knowledge available to the senses. His aim was to provide a mathematical description for the movements of the heavenly bodies, a model that could be used in predicting astronomical events. It was important to him that it not offend against the Christian belief system. That system, insofar as it was interested in astronomy, had its roots in Pythagoras and Plato. Accordingly, the heavenly bodies were seen to represent the ideal of perfection and were therefore of perfect shape—that is, spherical—and subject to perfect, circular motion.

Ptolemy was quite explicit about accepting the metaphysical framework imposed by the philosophical world view that he subscribed to. In the *Almagest* he said, "We believe that the object which the astronomer must strive to achieve is this: to demonstrate that all the phenomena in the sky are produced by uniform and circular motions." And later: "Having set ourselves the task to prove that the apparent irregularities of the five planets, the Sun, and Moon can all be represented by means of uniform circular motions, because only such motions are appropriate to their divine nature . . . [w]e are entitled to regard the accomplishment of this task as the ultimate aim of mathematical science based on philosophy."

He saw the astronomer's job as finding mathematical models that would "save the appearances" of what was observed in nature —models that would account for observations of planetary motions that seemed to run counter to the accepted philosophical idea of the perfection of circularity. His success in this was so complete that his model dominated astronomy for fifteen hundred years. But it was far from a simple or intuitively obvious description. It was probably, for its time, as difficult as Einstein's General Relativity is in ours. Nevertheless, it worked—was made to work—for as long as the underlying belief system remained unchanged. It is worth noting that this is a characteristic of all successful scientific hypotheses, ancient and modern.

Ptolemy borrowed his central ideas of epicycles and Earth-centeredness from his predecessor Hipparchus of Nicea (c. 190–120 B.C.) Alternative, Sun-centered maps of the cosmos were not unknown in Hipparchus' time. Aristarchus of Samos (c. 280 B.C.) explored a model that had the planets orbiting the Sun but still had the Sun circling the Earth. He rejected that in favor of the more straightforward plan of all of the planets, including Earth, orbiting the central Sun. He recognized that this would mean that the sphere of the fixed stars was unimaginably distant from Earth, so far removed that, seen from one of the fixed stars, the Earth's orbit would be as a single dot. This, he presumed, was the only way to account for the fact that no shift in the position of the stars could be detected even though the Earth traveled great distances in its annual orbit around the Sun.* We know this from the writings

*That is, he could detect no stellar parallax, a phenomenon that will be discussed later.

of the great Archimedes (287–212 B.C.), to whom Galileo would acknowledge an enormous debt for showing him how to approach the laws of motion, and whom Newton thanked for presenting him with the notion of centers of gravity. His prodigious gifts as a mathematician and physicist notwithstanding, Archimedes dismissed the Sun-centered system as physically preposterous and aesthetically monstrous. His views were shared by virtually all astronomers of late Antiquity—indeed, all educated people—right down to the time of Copernicus.

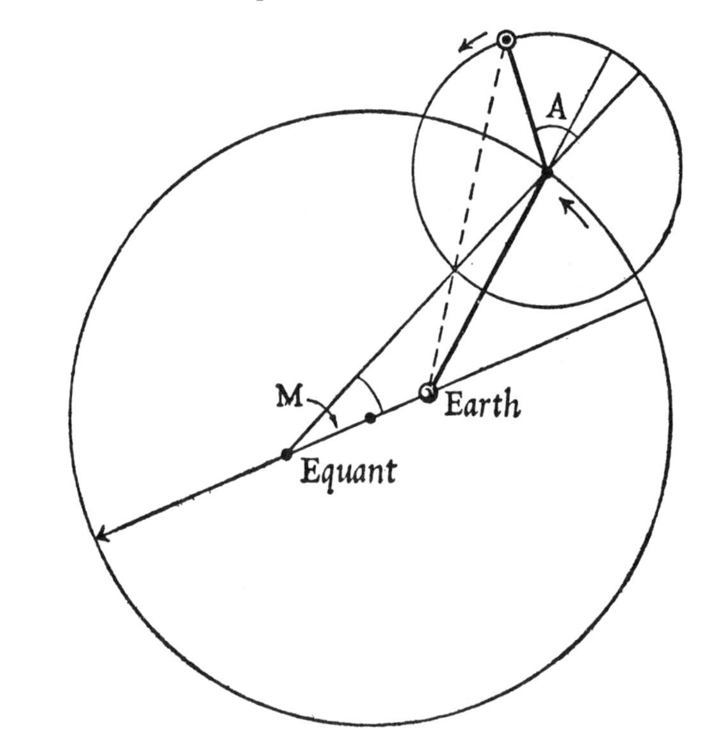

Ptolemy's construction of planetary motion. Angles M and A increase uniformly but differently for each planet. For Mars, Jupiter, and Saturn, M follows the planetary period and A makes one revolution a year. For the inner planets, Venus and Mercury, the reverse is the case. The dotted line shows the planet as seen from Earth. The further complication of the circular equant orbit (see text) is not shown.

Ptolemy adopted from Hipparchus the key idea that the sometimes erratic motions of objects in the night skies could be accounted for accurately by assuming that though all celestial motion was circular, some circular orbits, like that of the Moon, were centered on the Earth while others, those of the planets, were centered on points along a series of concentric circles drawn around the Earth. These latter orbits are called epicycles. Anyone who has been to a big carnival midway will have seen epicycles in action in a ride in which two small Ferris wheels are attached to the ends of a long spoke with a hub at its center. As the spoke rotates on its hub, the Ferris wheels at the ends of the spoke also rotate, describing epicycles in the sky. If the Earth is imagined as the hub of the long spoke, the wandering planets in their motions follow both the big circle described by the tips of the spoke, and the little circles made by the outer Ferris wheels as they turn on their own hubs. This nicely accounts for the observed back-and-forth wanderings of the planets, since for half of each epicyclic orbit, planets will be moving "backwards" in relation to the overall motion of the spoke.

Ptolemy added the complication of eccentric orbits. This was to account for the observation that the planets appeared to move closer and farther from Earth, and to speed up and slow down in their orbital motion. The eccentric orbit was modeled this way: the center of a given planet's orbit around Earth was not Earth itself but a point called the equant, which was located in space near the Earth. Around the equant Ptolemy drew a small circle, and the points along this circle provided a moving center to the orbit of the planet. The planet orbits in a perfect circle around a point that itself moves in a small, perfect circle. The result is a planetary motion around Earth that is actually eccentric or egg-shaped. This accounts neatly for the fact that planets appear at some times of the year to be farther from Earth than at others, but it also confers on them an apparently uniform velocity throughout their orbit since the position of the equant could be moved to make the math correspond to what was actually observed in the night skies. Thus what seemed to be a speeding up and slowing down of motion that *ought* to be perfectly uniform could be shown to be an illusion.

This may at first glance seem a trivial, if complicated, exercise in mathematical sleight-of-hand, but it is far more than that. Like all mathematical models that carefully match conditions in the real world, it became an extremely powerful tool. It made it possible to predict future astronomical events, and recreate events of the past, conferring the power of virtual time travel on those who understood its subtleties.

When Galileo was a child of eight living in Pisa, the first of two events occurred that shattered the accepted conceptions of the universe. They made it impossible for astronomers any longer to believe in the solid crystalline spheres on which the planets rode, or in the idea of an eternally unchanging cosmos. In 1572 a new star appeared in the sky—a supernova—and for more than a year it was so bright as to be visible in daylight. This was clear evidence that the heavens were not immutable, as Aristotle and Ptolemy had claimed, but were subject to change. And then in 1577, when Galileo was a youth of thirteen, an exceptionally bright comet traced a path across the skies, following a course that the great Danish astronomer Tycho Brahe was able to determine was well beyond the Moon's orbit, possibly as remote as Venus. The comet's observed course would have taken it crashing through the supposedly crystalline spheres so lovingly described in Dante. Tycho was forced to conclude that the spheres were not solid at all, in fact probably did not exist. Historians have suggested that these events did more to shake the foundations of the belief in the Ptolemaic-Aristotelian cosmos than the publication of Copernicus' hypothesis. Certainly they would have been seen, and remembered, by Galileo. A radical theory began gaining adherents: it proposed that the planets swam independently in a transparent fluid akin to air.

Three more comets were observed in 1618, and the phenomenon's ancient reputation as a harbinger of earthly disaster seemed to be borne out. This was the year of the outbreak of the catastrophic Thirty Years War, the first general European conflict, fought mainly on the soil of the present-day Czech Republic and Germany by barbarous mercenary armies.

The extraordinary, inexplicable appearances of the supernova and comets fueled the already lively revival of interest in astrology that had been a feature of the Renaissance. Astrology had been taught and practiced by the most admired of the major Hellenistic philosophers. It is thought that they, in turn, learned it from the Babylonians in the time of Alexander the Great's eastern conquests. It was a system of knowledge of extreme antiquity—the records of Babylonian priests went back at least as far as the third millennium B.C. and professed to stretch back thousands of years before that. It was of course the precursor to the science of astronomy, and astrology and astronomy continued to be practiced in tandem right up to the time of Newton, when the increasing formalization of science forced an irreconcilable split. Tycho was an acknowledged master of astrology; Galileo himself was a dabbler, though by no means an adept. In 1609 he cast a horoscope for the Grand Duke Ferdinand I of Tuscany, foretelling a long and happy life. The duke died shortly thereafter, and Galileo's student, Prince Cosimo, became the new grand duke.

But there is no better example of the concurrent interest in these cognate fields than Johannes Kepler (1571–1630), who had been a student of Tycho Brahe's and was the heir to a mountain of his teacher's very accurate but chaotically disorganized astronomical records. Kepler also inherited Tycho's position as imperial mathematician to the Holy Roman Empire.

Although, as we'll see, astrology informed his entire career, Kepler's reputation as an astronomer derives from his extraordinarily detailed and disciplined ten-year study of Tycho's data, from which he arrived at a conclusion that transformed astronomy—though its importance was little recognized in his time, least of all by his long-time correspondent Galileo. The conclusion was that the orbits of the planets are not circular but elliptical, and the Sun is not at the center, but at the focus, of the ellipses. With this discovery, the need for epicycles and eccentrics and all the arcane geometrical bric-a-brac that had cluttered astronomy simply vanished. He published his momentous conclusions in 1609, in a work called *The New Astronomy*. But it was not until eighty years later when Newton was able to demonstrate that Kepler's findings could be mathematically deduced from the law of universal gravitation that the (by then) long-dead astronomer's genius was fully appreciated. Indeed, Kepler seems to have been the finest mathematician and theoretical astronomer of his age, the superior of Galileo in both fields. He correctly guessed the Moon's influence in causing tides, and seeing from Tycho's records that planets move more quickly when they are in the vicinity of the Sun he developed a theory from which it is only a small step to Newton's law of universal gravitation. Kepler was far ahead of either Copernicus or Galileo in lifting astronomy out of its Aristotelian rut and placing it on the road of modernism, by wedding it to mathematics. He aspired to turn the universe into pure clockwork. "My aim," he said in a letter, "is to show that the heavenly machine is not a kind of divine, living being, but a kind of clockwork (and he who believes that a clock has a soul, attributes the maker's glory to the work), insofar as nearly all the manifold motions are caused by a most simple, magnetic and material force, just as all motions of the clock are caused by a simple weight."

And in another he said that humanity will correctly assess its powers only when it understands "that God, Who founded everything in the world according to the norm of quantity, also gave man a mind that can grasp these norms. For like the eye for color, the ear for sounds, so man's mind is not meant for the knowledge of whatever arbitrarily chosen things, but for that of magnitudes; it understands something the more correctly, the more it approaches pure quantities, as the origin of the thing."[1]

But if Kepler was, like Galileo, a prototypic modern mechanical realist, Kepler had other attributes that were not shared by his more famous contemporary and that science would soon come to regard as embarrassments. For the nonscientist they are what makes his career so extraordinarily fascinating and instructive. He was, for one thing, an unabashed disciple of the mystic Pythagoras and held a literal belief in the music of the spheres. He pursued his career not in the disinterested quest for scientific fact but while passionately seeking evidence of mystical numerical and geometric relationships between celestial bodies.

It was in that cause, and by no means as a detached scientific observer, that Kepler made the three momentous discoveries that he is best remembered for. First, as we've seen, he found that the focus of planetary orbits is not an abstract point in space near the Sun but the Sun itself, and the orbits of the planets were not circular but elliptical. Second, he discovered that if a line is drawn from an orbiting planet to the Sun, that line sweeps over equal areas in equal times, and that the squares of the period of the planets' orbits are proportional to the cubes of their mean distances from the Sun. This insight neatly explained the fact that the planets were observed to be moving faster when they were closest to the Sun, and slower when they reached the extremity of their orbits. Third, he found that a definite ratio exists between the size of the planets' orbits and the time it takes them to complete a revolution around the Sun.

Each of these insights would be incorporated into Newton's

modern system of dynamics. But for Kepler they were not impersonal laws, and the universe in which they functioned was not the chance product of their blind operation. His was a cosmos of consciously contrived order and harmony. "The creator," he said, "does nothing by chance."[2] Where resources in nature are adapted to specific ends, "there order exists, not chance; there is pure mind and pure Reason."*[3]

Kepler's science was deeply Christian as well as theistic. He criticized Ptolemy for his pagan belief that the stars were visible gods, an idea, he said, "that cannot be tolerated in a Christian discipline."[4] Wherever he looked in the universe he saw things organized in triads, mirroring the Christian Holy Trinity. His universe was finite and spherical because the sphere represented to him the embodiment of the Trinity, with the center symbolic of the Father, the surface the Son, and the intervening space the Holy Ghost. He calculated the densities of the Sun, the sphere of the fixed stars, and the intermediate ether by assuming that the three persons of the Trinity demanded equality. The sphere, the emanation of a single point, was to Kepler the image of the divine, the perfect shape toward which everything aspires, to the extent it is able to do so. Bodies that are confined within various shapes nevertheless expand their presences spherically by means of their senses and powers.

And he could further intensify the Trinitarian imagery: "A point, the center of a sphere, is invisible; it reveals itself by flowing outward in all directions. The surface is its image, the way to the center. Who sees the surface also sees the center, and in no other way."[5]

Kepler worked out a fantastic cosmology based on geometrical figures nested within one another. Each of his astronomical discoveries only served to further confirm him in his Christian and Pythagorean mysticism, which evolved, as it had for the original Pythagoreans, into a belief in the reality of pure number. Kepler

---

*Nor was Newton himself the paragon of scientific objectivity he is often made out to be. Recent examination of Newton's "scholia," his personal commentaries on his published writings, has revealed that he evoked the Pythagorean idea of the music of the spheres and its mathematical basis in order to explain his formula for universal gravitation. Newton's personal notes have also revealed a secret interest in alchemy.

saw mathematics as the tangible link between man and God: "Why waste words? Geometry existed before the Creation, is coeternal with the mind of God, *is God himself* (what exists in God that is not God himself?); geometry provided God with a model for the Creation and was implanted into man, together with God's own likeness—and not merely conveyed to his mind through the eyes."[6]

And even more distressing for the modern scientist than his mystical and religious convictions and their intrusions into his science is Kepler's stubborn and profound belief in astrology. He began his career with the publication of astrological calendars and ended it as court astrologer to the duke of Wallenstein. He wrote a number of treatises on the subject, and it creeps continually into his scientific works. He sought to cleanse astrology of its encrustations of superstition and urged theologians, philosophers, and scientists "while justly rejecting the stargazers' superstitions, they should not throw out the baby with the bath water . . . [since] nothing exists nor happens in the visible sky that is not sensed in some hidden manner by the faculties of Earth and Nature: [so that] these faculties of the spirit here on Earth are as much affected as the sky itself."[7] That the planetary realm influences humankind was perfectly obvious: "The belief in the effect of the constellations derives in the first place from experience, which is so convincing that it can be denied only by people who have not examined it." Even though "what it does specifically remains hidden,"[8] it was only logical to assume that human fate was determined by the same celestial influences that shaped the weather and the seasons and determined the fecundity of herds and the richness of harvests.

The journalist and novelist Arthur Koestler (1905–83) has suggested that in Kepler's astrological determinism can be found the antecedent for the biological and psychological determinism of twentieth-century human sciences. Most scientists would vehemently reject the notion, but it nevertheless has an irresistible appeal. Kepler's determinism was neither more nor less irrational than modern scientific reductionism, as exhibited, for example, in the discipline of sociobiology. What, after all, is there to distinguish among any group of deterministic theories, once it is understood that they are all ultimately unprovable? Only that some seem to explain and predict lived experience in a more satisfying and useful way than the others do. Kepler, though a scientist, was convinced on the balance of the evidence of lived experience that astrology was a valid hypothesis, that the experiences forecast by its general principles did occur as predicted.

Kepler's fruitful career also demonstrates that not just scientific but religious and mystical hypotheses as well can be profitable in exploring the world and developing instrumentally useful knowledge. His experience highlights the fact that when useful knowledge is derived by inquiry inspired by a given hypothesis or methodology it is by no means safe to assume that this demonstrates the truth of the hypothesis—it testifies only to its utility. True conclusions, as Aristotle knew, can be derived from false premises. This is of course as correct of scientific hypotheses as religious ones. Kepler moved astronomy forward by giant strides—greater than Galileo's—by thinking within a framework that modern science finds completely unacceptable. There is no reason in logic to assume that similar progress could not be made today on the basis of similarly "misguided" mystical and spiritual insight. Of course, in today's environment, the attempt could not be made, so complete is the domination of science and its culture by pure reason. In nothing so much as the studied contempt in which he would hold Kepler's astronomical discoveries is Galileo entitled to be called the first modern scientist.

FIVE

Young Galileo • The Studies in Motion
The Experimental Method
Number and Beauty
The Pythagoreans and the Reduction of Quality to Quantity

Young Galileo attended school at a Benedictine monastery near Florence, where his family had moved in his tenth year. He must have found the atmosphere congenial, because he joined the order as a novice, intent on becoming a monk. His father would not allow that to happen—it would have been a continuing financial burden the family could ill afford. He sent him instead to the University of Pisa to study medicine, which was a remunerative trade. In 1583 Galileo began receiving private tutoring in mathematics from Ostilio Ricci. Ricci was a professor at the Florentine Academy of Design, where he taught a number of subjects including mathematics, perspective, astronomy, mechanics, architecture, and anatomy. He was also mathematician to the Tuscan court. It was the custom at the time for the court to move from Florence to Pisa between Christmas and Easter, and it is during that time that Galileo joined the court pages for Ricci's lectures on Euclid. He was, by all accounts, enthralled. The young man's aptitude, and appetite, were so remarkable that Ricci, a family friend, was able to persuade Vincenzo to allow his son to forgo studies in medicine and pursue his mathematical talent. Ricci continued to tutor Galileo in Euclid and Archimedes.

There is a famous story that Galileo, during his second year of university, was watching a lamp swinging in the glorious Romanesque cathedral in Pisa one morning. (He was, we are to presume, bored with the cant and ritual of the service.) It suddenly occurred to him that the lamp always required the same amount of time to complete an oscillation, no matter how wide the swing. This insight led him to suggest the pendulum as a regulating mechanism for clocks. Even today, guides at the cathedral make an ancient bronze lamp hanging in the nave a feature of their tours. The story as it is usually told is a thinly veiled allegory highlighting the superiority of the scientific mind. While the rest of the congregation wasted their time in the protocols of religion, Galileo's scientific mind was alert to the truth. He did in fact experiment with the pendulum later in life, and suggested its use in timekeeping (though he was not the first to do so). But unfortunately for the story, the lamp that he is supposed to have been watching as a youth was cast in 1587, by which time his student days had ended.

A lack of money—his perpetual bête noire—forced Galileo to withdraw from the university in his fourth year, before he had been granted a degree. He returned to Florence in 1585 to continue his studies privately and to eke out a living tutoring mathematics in Florence and nearby Siena. In 1587 he was able to journey to Rome to visit one of the leading mathematicians of the age, the German Jesuit Christopher Clavius, and the two continued to correspond for many years.

To this period belong two lectures on the size and shape of hell in Dante's *Inferno*, delivered by the young scholar before the Florentine Academy in 1588, the year in which Spain's mighty Armada was destroyed off Britain's coasts, and the philosopher Thomas Hobbes was born. During this period as well, Galileo invented a hydrostatic balance for determining the specific gravity of objects, and wrote a small treatise on it called *The Little Balance*. The treatise highlights his high regard for Archimedes, the inventor of the science of mechanics, whose famous "Eureka, I have it!" introduced to the world the concept of specific gravity. It was Archimedes' ability to translate his observations of mechanical phenomena such as the power of levers into mathematical relationships that most impressed Galileo. Inspired by Archimedes, the young scholar worked out several theorems on specific gravity, which he did not publish until much later, in his *Discourses on Two New Sciences* (1638). The papers were, however, circulated privately, and they earned Galileo the respect of many of the mathematicians of his time.

Galileo applied to several universities, looking for a teaching position, and was finally offered the chair of mathematics at his alma mater in Pisa, thanks to the impression he had made on the Tuscan court with his Dante lectures and to recommendations from Christopher Clavius and other mathematicians. He returned there in 1589 on a three-year contract. It was not a high-status position and the pay was poor—just sixty scudi a year compared with the two thousand scudi paid to a professor of medicine. However, as an Italian biographer puts it, "Galileo was able to reenter with the title of professor the same university he had left four years earlier without a degree. His personal honor was vindicated."[1]

It was at the University of Pisa that Galileo was introduced to the work of Giovanni Battista Benedetti. Benedetti was an exponent of the ancient theory of impetus, which had its roots in the sixth century in Philoponus, who used it to challenge Aristotle's concept of "violent" or projectile (as opposed to "natural") motion. Aristotle had seen the cause of the continued motion of a projectile as being in a force exerted by the medium through which an object was moving. But Philoponus had theorized that something called *impetus* was transferred from the source of propulsion to the object being propelled, and thereafter was contained in the moving object. Benedetti was also a Copernican. Although Galileo was then dutifully teaching courses in Ptolemaic astronomy to his students, it is generally thought that he began privately to entertain the Copernican hypothesis at this time.

His studies into motion began, and his researches are preserved in a collection of unpublished papers he labeled "Older Studies of Motion." It is interesting to note that in his inquiry into the motion of falling bodies in this period, he initially believed that the speed of fall was related to the density of the object, so that while two balls of lead of differing size would fall at the same rate, a ball of iron of any size would fall more slowly. It was not until much later that he concluded that all objects, regardless of weight or density, fall at the same rate, and this finding was published for the first time in his *Two New Sciences*.[2] For this reason most scholars dismiss as pure invention the story reported in the early biography by his adoring student Vincenzo Viviani, of the experiment at the Leaning Tower of Pisa. According to this account (published fifty years after Galileo's death) the scientist is supposed to have simultaneously dropped two balls of differing weight from the tower to demonstrate that they would hit the ground at the same time. When they did, the story goes, the philosophers of the university, hidebound adherents of Aristotelianism, were confounded and modern science began.

While it is probably untrue that Galileo dropped objects from the tower to disprove Aristotle's physics, there is no doubting that he was convinced that Aristotle was wrong in stating that the rate of descent is proportional to the weight of the falling object. The idea was preposterous. "Try, if you can," he once urged an opponent, "to picture the large ball striking the ground while the small one is less than a yard from the top of the tower."

But he was far from original in his skepticism. Scholars throughout the Middle Ages had questioned Aristotelian doctrine on this point, and a famous mind experiment had been conducted to disprove it. It worked this way: imagine two balls of equal weight dropped simultaneously from a tower. Halfway down, an angel connects the two with a thin iron rod, so that there is now one object double the weight of either of the original balls. Would the connected balls suddenly fall faster than they had while unconnected? Of course not, was the answer.

There were other more or less obvious errors in Aristotle's physics that had long been appreciated. For example, Aristotle had said that a projectile follows a straight path until the impetus driving it forward is exhausted, and then falls vertically to the ground. It would have been difficult to find a military man in any age who would subscribe to this. Certainly the designers of the military catapults of the Middle Ages, who were usually senior churchmen, would have known that their stones did not fall vertically at the end of their trajectory, as would the English archers whose marksmanship took such a heavy toll on the French armies in the time of Richard the Lion Heart. In 1546 a Venetian military engineer named Niccolò Tartaglia had published a book on artillery tactics and munitions in which he made note of the fact that both gravity and the force of impetus act on a projectile throughout its flight, so that it follows a curved trajectory. He wrote that the optimal angle of elevation for long-range artillery was forty-five degrees, and that at angles greater or lesser than this the range falls off, at first slowly and then rapidly.

What Tartaglia had observed as a fact, Galileo set out to prove experimentally during his years as a teacher in Pisa. Using polished wooden ramps and metal balls that rolled down them at speeds he could measure more or less accurately, he eventually demonstrated that all bodies, irrespective of their weight, fall through the same distance in the same time. He also discovered that the velocities of falling bodies increase uniformly over time (the distance of the fall being proportional to the square of the time taken in the descent). It was during these years that Galileo brought the mathematical-experimental method of science to maturity, though again, it would be many years before he published his findings.

The enormous significance of the mathematical-experimental method is perhaps not immediately obvious. Ptolemy and others had devised mathematical models to account for natural phenomena such as the movements of the planets. The brilliant Flemish contemporary of Galileo's, Simon Stevin, had conducted experiments to confirm mechanical hypotheses. But Galileo combined the two ideas and that allowed the scientist, by carefully observing experiments, to build mathematical models that were universal in application. That meant that natural phenomena could be predicted even before they had been observed. Prior to this, new phenomena had only been found by chance or inspired guess. The experimental-mathematical approach systematized the scientific enterprise and greatly accelerated its progress.

Galileo was a reductionist not only in his science, in which he insisted on a separation of quantity from qualities, but in his critical appreciation of art. "Music," he said, "if we wish to penetrate to the essence of its being, must be taken as instrumental music, detached from words and, above all, from dramatic representation."[3] His strong theoretical bias is also evident in his approach to media. In musings on the relative merits of painting and sculpture he comes down on the side of painting, in part because "the farther removed the means of imitation are from the thing to be imitated, the more worthy of admiration the imitation will be."[4]

If Galileo's comments on music and painting may be taken as a guide, we might expect him to have approved of simpler designs that followed more transparently the rules set down in Leon Battista Alberti's influential *Treatise on Architecture*. Alberti (1404–72) insisted that the arithmetical ratios that determine musical harmony must also govern architecture, because they had been found to recur throughout the universe.

In Alberti's approach to architecture, beauty was "reduced" to geometry.* In Galileo's science, all nature would be so reduced. Both men shared a devotion to the power and beauty of numbers that arose out of a two-thousand-year-old philosophy that had been enthusiastically resuscitated early in the Renaissance.

Although it would come to be accepted as the epitome of scientific rationalism, the idea of reducing nature and experience to number arose out of the insights of one of history's greatest mystics, Pythagoras of Samos. In Greek tradition Pythagoras is part magician and part mathematician. He is regarded as the first to carry an interest in mathematics beyond the requirements of commerce, to study it for its own sake. Between about 529 and 509 B.C., after traveling to Egypt and Persia and learning their mathematics and mystical lore, he set up a series of cult-like schools in the town of Croton, which is on the southern lip of the Gulf of Taranto, on the sole of the boot of Italy. In time, he and his followers came to dominate the political life of the city. Their schools were run like secret societies, and the members were required to lead an ascetic life that

---

*One could argue that *elevated* might be a more appropriate term than *reduced*. I use the word *reduced* here in its technical philosophical context of *reductionism*, which is the notion that the social sciences can be reduced to physics, or that all natural phenomena that have a real existence can ultimately be described in mathematical formulas.

may have provided a model for early Christian monasticism.

The Pythagoreans' fascination with number in nature is said to have begun with the simple observation that the pitch of a musical note depends on the length of the string that is vibrating to produce it, and that the concordant intervals in the musical scale are produced by simple numerical ratios. Halving the length of a string produces a difference in pitch of an octave; two-thirds of the length a fifth; and three-quarters of the length a fourth. It was presumed that if these same ratios—2:1, 3:2, and 4:3—were applied to voids and solids, a visual harmony analogous to the musical would result. This was the direct source of Alberti's theory of design, as well as of that of Andrea Palladio, perhaps the quintessential Renaissance architect.

Whether it was music or some other phenomenon that provoked the Pythagoreans' obsessive interest in numbers matters little. What is of profound historical significance is where the original insight led them. What the Pythagoreans had found—and what was to be of intense interest to Galileo and other early architects of science—was a way to reduce *quality* to *quantity*. They had found a way to reduce the human experience of nature, even nature itself, to number. They raised for the first time the possibility and even the necessity of providing a quantitative explanation of the cosmos. They invented reductionism and made mathematical, mechanistic science possible.

For the Pythagoreans, the universe was composed of numbers, and it was exclusively through numbers that knowledge could be won. "And in fact all things that can be known have a number," says a surviving text, "for it is not possible for anything either to be thought or to be known without this. . . . For the nature of number is the cause of recognition, able to give guidance and teaching to every man in what is puzzling and unknown. For none of the existing things would be clear to anyone unless there existed Number and its essence." Aristotle reports in his *Metaphysics* that "the so-called Pythagoreans, who were the first to take up mathematics, not only advanced this study, but also having been brought up in it they thought that its principles were the principles of all things. . . . In numbers they saw many resemblances to the things that exist and are coming into being—one modification of number being Justice, another Reason, another Opportunity—almost all things being numerically expressible. Again they regarded the attributes and ratios of the musical scale as expressible in number. They therefore regarded numbers as the elements of all things, and the whole heaven as musical and numerical scale."[5]

Because only fragments of their writings have survived, it is not entirely clear whether the Pythagoreans believed that number was a contingent property of things, or the essential nature of things—whether number happened to be *in* things or was the things themselves. It may be that they began by believing the former and gradually came to be persuaded of the latter. We do know that Pythagoreans believed that the distances of the various heavenly bodies from the Earth corresponded to musical intervals. The harmonies created by their motion were actually audible to the pure of heart—the famous "music of the spheres." They explained that the reason we are not normally conscious of it is that we are immersed in it from birth and have no experience of a world in which it is not present.

There is, in spite of the inherent coldness of mathematics, a compelling beauty in the Pythagorean conception of number and harmony in nature that has been appreciated by poets and artists down through the ages. In the book of Job, God speaks out of the whirlwind of Creation as a time "When the morning stars sang together, and all the sons of God shouted for joy."

Dryden wrote:

From harmony, from heavenly harmony,
This universal frame began:
When jarring nature underneath a heap
Of jarring atoms lay
And could not heave her head.
The tuneful voice we heard from high:
Arise, ye more than dead.

And Shakespeare, in *The Merchant of Venice*, speaks of how

. . . soft stillness and the night
Become the touches of sweet harmony . . .

There's not the smallest orb which thou behold'st
But in his motion like an angel sings,
Still quiring to the young-eyed cherubins;
Such harmony is in immortal souls;
But, whilst this muddy vesture of decay
Doth grossly close it in, we cannot hear it.

The pioneering work of the Pythagoreans is clearly evident in the second-century A.D. astronomy of Ptolemy (who wrote a three-volume treatise on music called *Harmonica*). It had been preserved by Plato, who enjoyed a close association with at least one disciple of Pythagoras'. Through Plato, the fascination with number was passed on to Aristotle. And through Plato and Aristotle it profoundly influenced Christian theology and philosophy. Revived in the Renaissance, it was imbued with fresh significance by Copernicus, Kepler, and Galileo. It lives on in Newton and the physicists of our own era.

As for the original cult of Pythagoras, it came to a sad and ironic end. The brotherhood was severely shaken by the discovery that not all quantities can be expressed in ordinary whole numbers —that some quantities can only be expressed in terms of so-called irrational numbers. For example, a line drawn diagonally across a square whose sides have a precise numerical length cannot be described in terms of a real number—it will have a fractional value that can only be an approximation and as a decimal will extend itself indefinitely. The square root of two is another example. Pi is a third. Such numbers are neither odd nor even, and there was no place within the cut-and-dried Pythagorean cosmos for such monstrosities, which they called *arrhetos*, the unspeakable. What was so shattering was that these irrational numbers—which can hardly be called numbers at all—are not exceptional but common.

Just as the promulgating of Heisenberg's Uncertainty Principle and Gödel's Theorem were to do twenty-five hundred years later, the discovery of irrationals destroyed any hope of uncovering universal verities through the application of mathematics. In each case, mathematics had experienced the limits of its ability to express the nature of reality, exposing a chasm between mathematics and reality that is unbridgeable by pure reason.

According to tradition, the Pythagoreans kept the scandal of their discovery a closely guarded secret, and the initiate who eventually leaked the news was drowned by his colleagues. More reliable historical accounts describe how Pythagoras and his followers were driven out of Croton by an uprising of the citizenry. Pythagoras' political ideas were tainted by the same authoritarianism found later in Plato, and he was as severely ascetic a leader as any Cathar priest. Citizenry led by the democratic faction of the city set fire to a house where Pythagoras and his followers were meeting and drove the survivors out of town. Pythagoras died in exile. It is sometimes said he starved himself to death.

SIX

Ferrara and Copernicus
The Reluctant Revolutionary
Challenging Aristotle • A Stimulating Meeting

I set out one weekend in the direction of Venice, where Galileo introduced the telescope and where he set his *Dialogue on the Two Chief World Systems*. On the way, I made a stopover in Ferrara, where I had arranged through contacts in Toronto to meet a specialist in the science of the Renaissance. Apart from her academic credentials I knew only that she was a Dominican nun and that she lived in Venice but spent much of her time at universities in Ferrara and Rome. She was called Sister Maria Celeste—Sister Celeste for short—and we were to meet the following day near the entrance to the city's famous cathedral.

I parked my rental car in the city's central square near the twelfth-century Duomo with its intricately detailed marble facade, a hybrid of the Gothic and Romanesque, a riot of columns, niches, windows, and finials. Along the south side of the cathedral are permanent stalls for a farmers' market that has been operating here since the fifteenth century. Nothing, I thought, could more aptly illustrate the way that the Church had been integral to the fabric of life in those days than this magnificent cathedral with its built-in commercial loggia. One can easily imagine the market's produce and livestock and crafts and entertainers spilling over into the cathedral portico and filling the square, in a noisy, colorful, odoriferous hubbub. In the cathedral nave, men and women from all stations of life, along with their dogs, would have mingled to exchange gossip and negotiate transactions. Today that activity has moved out of the church into the secular space of the modern shops and arcades around the square, and the cathedral, except for Masses, is silent as a tomb, even on market days.

My gaze wandered to the red-brick wall of a building that adjoins the Duomo, where high up on the second story I noticed an engraved stone marker. The name Nicolaus Copernicus jumped out at me. The inscription was in Latin, but as nearly as I could guess it had something to do with a Copernican connection to the local university. I was to learn presently from Sister Celeste that Copernicus had indeed attended university in Ferrara, attracted by tuition fees that were lower than those of the big universities in Padua and Bologna. This sensible frugality was typical of a man whom modern biographers have found maddeningly overcautious in holding back news of his radical redesign of the cosmos until he was at death's door.

Copernicus (1473–1543) seems to have been one of those perpetual students found on university campuses in any age, piling degree upon degree, brilliant but unfocused and unsure of their talents, comfortable only in the rarefied company of scholars. He was thirty-three when he received his doctorate in canon law at Ferrara, and by then had studied astronomy at the University of Krakow near his home in eastern Poland, law and medicine at the University of Padua, and mathematics and the classics, as well as astronomy, at the University of Bologna. In all, he spent fourteen years in university. This extended career as a student was made possible by a feature of the economy of the time, the well-paid sinecures available in large numbers to those with good connections within the Church. These were of course financed by the tithes and taxes levied on ordinary citizens. Through the influence of his uncle, the bishop of Ermeland, Nicholas, at twenty-eight, was appointed a canon of Frauenburg, a town now in northern Poland but then part of the Prussian Estates. This was an administrative rather than a confessional position—Copernicus was not a priest. But it guaranteed lifelong financial security in return for a minimal commitment of time and energy. Galileo himself benefited from this system of prebends, holding the title of canon of both Brescia and Pisa in his later life. Although he was not required to wear ecclesiastical garb, or indeed even to visit the cities in question, he did have his head shaved in a bald-pated ecclesiastical tonsure by the bishop of Florence.

In the course of his long studies, Copernicus had become increasingly dubious about the validity of the Ptolemaic system. His skepticism was not based on any special insight but rather on a punctilious personality's annoyance with a feature of the Ptolemaic system that seemed unnecessarily untidy. Copernicus, along with everybody else, took it as a given that motion in the heavens was always circular and always at uniform speed. Irregular motion could be conceived of on Earth, where the corrupt and the impure held sway, but certainly not in the sphere of the heavenly bodies, where perfection reigned. And yet, uniformity of motion in the Ptolemaic system was achieved only through the fiddle of the equant. So long as it was accepted that systems such as Ptolemy's had no purpose beyond providing a usable mathematical description and thereby "saving the phenomena," the only criteria for judging one system better than another were elegance and ease of use, which boiled down to simplicity. Copernicus existed on the cusp of an era when inquiring minds looking up at the heavens would push beyond simply saving the phenomena to seeking to understand the actual inner workings of physical reality. As a man of this transitional time Copernicus was seeking not just a mathematical formula but a physical explanation of planetary movements. But at the same time he was unable to jettison the Aristotelian dogma of perfection:

> Having become aware of these defects [i.e., the complexities of the Ptolemaic system], I often considered whether there could perhaps be found a more reasonable arrangement of circles, in which everything could move uniformly about its proper center, as the rule of absolute motion requires. . . . I therefore went to the trouble of reading anew the books of all philosophers on which I could lay hands to find out whether someone did not hold the opinion that there existed other motions of the heavenly bodies than assumed by those who taught

> the mathematical sciences in the schools. And thus I found first in Cicero that Hiketas has held the belief that the earth moves. Afterwards, I found in Plutarch that others have also held this opinion. . . . And so, taking occasion from this, I too began to think about the mobility of the earth, and although it seemed an absurd opinion, yet, because I knew that others before me had been granted the liberty of supposing whatever orbits they chose in order to demonstrate the phenomena of the stars, I considered that I too might well be allowed to try whether sounder demonstrations of the revolutions of the heavenly orbs might be discovered by supposing some motion of the earth.[1]

Copernicus was anything but a radical. His aim was not to revolutionize astronomy but to tidy up the work of his predecessors, Aristotle and Ptolemy. The fact that he had to put the Earth in motion in order to do this to his own satisfaction was almost incidental. The Copernican system was just the Ptolemaic system realigned and simplified. Its major advantage was that in refocusing the planetary system on the Sun it got rid of the messy geometrical problems raised by the retrograde or backward-and-forward motion of the planets, the motion that led the ancients to call them wandering stars. When you looked at planetary motion from the Copernican perspective, it became obvious that each time the Earth in its orbit overtakes one of the slower-moving outer planets, that planet will appear to pause and then reverse its direction of motion for a

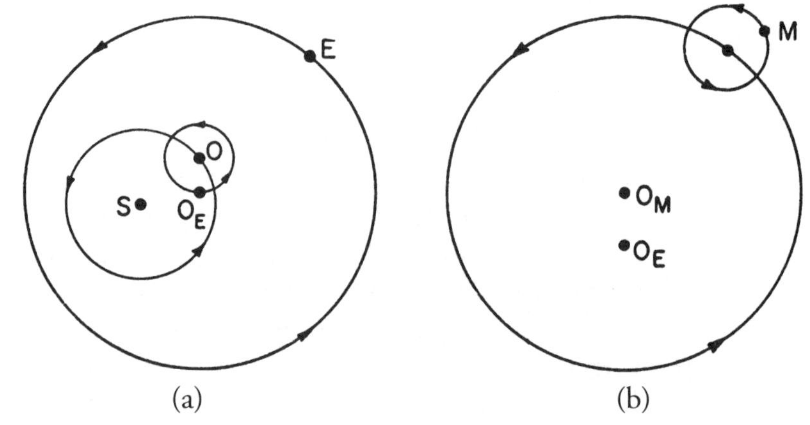

Copernicus' description of the motions of (a) Earth and (b) Mars. In (a) S represents the Sun; the Earth, E, revolves in a circle whose center is O_E, a point that in turn revolves around O. In (b) Mars is shown revolving on an epicycle that in turn revolves on a deferent whose center, O_M, has a fixed relationship to O_E, the moving center of Earth's orbit around the Sun. Despite claims for its simplicity, the system is as complex as Ptolemy's.

time. In the same way, each time one of the inner planets overtakes the Earth there is an apparent slowing and reversal of motion.

Despite this simplification Copernicus only slightly reduced the number of epicycles required to make the math of perfect circles conform with increasingly sophisticated planetary observations. And perhaps even more surprising is the fact that in the Copernican system the planets do not orbit the Sun at all—they orbit around a point in space near the Sun. In refusing to consider any orbital shape but the perfect circle (when in fact planets orbit in ellipses focused on the Sun), Copernicus had been forced to resort to most of the same geometrical expedients he had criticized in Ptolemy, in order to account for appearances. His system could be described as simply Ptolemy's with the Sun replacing the Earth at the center.

For all its inherent conservatism, and despite the fact that it was a description of the motions of heavenly bodies, the Copernican hypothesis as he elaborated in *De Revolutionibus—On the Revolutions of the Celestial Orbs* did raise immediate and pressing questions about earthly phenomena. This brought it into conflict with Aristotle, whose science and cosmology had achieved a status close to dogma within the Church in the 368 years since the death of his great advocate and interpreter, St. Thomas Aquinas. The first and most pressing conflict concerned the nature of gravity. Aristotle had defined gravity as the desire of all matter to return to its "natural" place at the center of the universe. If the Earth no longer held that position, how did one account for the fact that objects fell toward its center? Copernicus' response was that matter has a natural tendency to want to form itself into spherical arrangements—the sphere being the "perfect" shape. He remained Aristotelian at least in spirit. And though it may seem a charmingly quaint explanation, it should be noted that despite Newton and Einstein, science in the succeeding 450 years has been unable to provide a better one.

There were other common-sense Aristotelian objections to a moving Earth, to which Copernicus had Aristotelian replies. If the Earth moved, it was argued, objects dropped from a height ought to hit the ground not directly below the release point but some distance away. Common sense said that this was because the Earth would have moved, if only slightly, in the interval between release and impact. In fact, the thinking went, any object not firmly attached to the surface of a moving Earth—including the atmosphere—would get left behind. The Earth itself would fly to pieces if subjected to the terrific speed of motion suggested by a Sun-centered system. (Rotational speed on the Earth's surface at the equator was calculated to be about 1,000 miles an hour.) But Copernicus noted that Aristotle had carefully distinguished between "violent" motion and "natural" motion, and, the astronomer pointed out, it is in the nature of a sphere to rotate. "If one holds that the earth moves, he will also say that this motion is natural, not violent. Things which happen according to nature produce the opposite effects to those due to force. Things subjected to violence or force will disintegrate and cannot subsist for long. But whatever happens by nature is done appropriately and preserves things in their best conditions. Idle, therefore, is Ptolemy's fear that the earth and everything on it would be disintegrated by rotation which is an act of nature, entirely different from an artificial act or anything contrived by human ingenuity. . . ."[2]

As to the falling-object problem, he proposed another solution that Aristotle might have approved. In the Aristotelian cosmos, matter was composed of earth, water, air, and fire in various combinations. Since objects in the lower atmosphere where people dwell are relatively watery and earthy in the Aristotelian scheme of things (as opposed to airy and fiery), it was only natural to expect them to conform to the motions of the Earth, "to participate in the nature of the whole to which they belong." In other words, falling objects close to the Earth's surface participated in the Earth's own motion because they "belonged" to the Earth. Once again, what seems a slightly absurd explanation was actually a perfectly adequate hypothesis for its time. Galileo and his successors would conduct experiments demonstrating conclusively that objects on the Earth's surface do indeed share its motion, just as objects on a ship share the ship's motion, so that when a passenger jumps in the air, he lands on the same spot on the deck and not several inches astern. But none would have a better answer for the question, Why?, and it soon stopped getting asked at all in scientific circles. Galileo would not "discover" the laws of motion; he would invent them, just as Newton would invent the law of gravity. They were successful "laws" because they seemed to conform to observed experience and made possible predictions based on hypothetical conditions, predictions that could be tested in experiments. In that narrow sense these laws of science could be said to represent truth. But neither Galileo nor Einstein would come closer to explaining why nature behaves in the way it does than had Aristotle, who sought truth on a wider horizon.*

Sister Maria Celeste arrived exactly at eight, the only nun in sight. Because she had to walk right across the cobbled cathedral square to reach my table, I was able to observe her carefully as she approached. She wore the crisp black-and-white habit of the Dominicans, the skirt abbreviated to calf length. Her shoes were, of course, black and undecorated, but they were an exquisite quality that one can find in the best Italian shops. Her black briefcase, though it showed signs of heavy use, was of the finest Moroccan leather and the product of superb Florentine craftsmanship. She looked, in short, stylish in a quintessentially Italian way, despite her uniform. She had the kind of alert, intelligent, and mobile face that is always attractive no matter what its deviations from the classical norms of beauty may be, though I must add, there were few of these. As she approached with the purposeful stride of the nontourist, I judged her to be in her mid-forties and fit. So perfectly proportioned was she, and so well did her habit fit, that it was a surprise to realize that she was only about five feet tall, perhaps even slightly less, a feature of her physical makeup that was noticeable only when she stood next to a group of schoolchildren, waiting for traffic to clear so she could reach the terrace where I was waiting.

I stood to greet her, lifting the straw hat I had told her I'd be wearing, and she smiled and put out her hand; I took it in mine. She had only a little time before she had to catch the train to Venice, she said, and she suggested we find a table at a quiet bar up the street across

---

*There are, of course, different levels of "why" questions. Why do planets orbit the Sun? Because of the universal law of gravity. Why is there a gravitational attraction? Because of the structure of space. Why is space so structured? etc. At some stage, however, further explanation must cease. It can end because the limits of scientific knowledge have been reached, or because a definitive metaphysical answer of the kind Aristotle proposes is available.

from the massively fortified Este Castle with its moats and turrets. From our table we could see the big, brooding statue to the martyred Dominican monk Savonarola, a native son, thundering his prophecies of retribution for the pre-Reformation princes of the Church.

In an e-mail before leaving home I'd told Sister Celeste that I was especially interested in learning more about the Church's attitude to science in the seventeenth century. And now with only scant attention to formalities she began to extemporize on the subject, speaking in heavily accented but near flawlessly grammatical English. But she immediately interrupted herself to retrieve a newspaper from her elegant briefcase—that day's *Herald Tribune*. By what seemed to me an amazing coincidence, Pope John Paul II had been in Torun, Poland, on the previous day, and she wanted to show me the article. Torun is the birthplace of Copernicus, and in a speech to academics at the university there the pope defended the astronomer by quoting from his own recent encyclical on faith and reason. "As I wrote in *Fides et Ratio*," the pope said, "deprived of reason, faith has stressed feeling and experience, and so runs the risk of no longer being a universal proposition. It is an illusion to think that faith, tied to weak reasoning, might be more penetrating; on the contrary, faith then runs the grave risk of withering into myth or superstition." Copernicus himself, he said, "saw his discovery as giving rise to even greater amazement at the Creator of the world and the power of human reason." As he concluded his speech before the assembled professors and students, the paper reported, the pope exclaimed, "*Vivat academia! Vivant professores!*"[3]

"Of course," the little nun said, carefully folding the newspaper and replacing it in her bag, "he also spoke of the other side of the coin in the encyclical. He said that 'deprived of what Revelation offers, reason has taken side tracks which expose it to the danger of losing sight of its final goal.' This is all quite apropos of what you are curious about, yes? A question very old and at the same time very fresh, I think."

I nodded my agreement. "Aristotle meets Aquinas."

"The first thing you must understand is that it was not his debunking of Aristotle per se that got Galileo into hot water, it was those parts of Aristotle that were incorporated into Church dogma by St. Thomas Aquinas. But of course to understand Aquinas you have to go back to Aristotle. And you certainly must understand

Aristotle in order to understand the real causes of the dispute between Galileo and the Church. The Church's position on science was mainly St. Thomas Aquinas' interpretation of Aristotle, with a healthy dose of St. Augustine's Plato thrown in for good measure."

She smiled, glanced at her tiny stainless steel wristwatch, and reached back into her briefcase to retrieve a bundle of photocopied papers secured with an elastic band. She placed it on the small marble table between us, shifting the ashtray and her coffee cup to make room.

"These are some lecture notes which I hope someday, if I can ever find the time, to incorporate into an introductory text for students. You may find it helpful. If you would like, I can leave it with you and we can discuss it when we meet again. If you can possibly find the time to read it, I would recommend it. That way, when next we meet, we will be singing from the same hymn book, as they say."

She dipped into her briefcase a third time, finding a calling card, which she handed to me.

"This is my telephone number in Venezia. If you telephone, we can meet. If you take the train from Rome it is no problem, a few hours only."

She acknowledged my thanks with a smile and another glance at her watch and continued.

"It is important to know that Aristotle was a vitalist and a teleologist, which means—I am sure you know—that he believed that nature in its works proceeds toward a goal or *telos*, as the Greeks said, and that living entities are only partly controlled by their physical and chemical makeup, and are to varying degrees self-determining. He was, you could say, reacting against Democritus and the mechanistic philosophers of his era. They believed that the universe consisted of matter and motion and nothing else, that matter was completely passive, that the behavior of all living things was determined by the interactions of the atoms of which they were composed. We might today call them behaviorists or possibly evolutionary biologists, and they certainly have the sympathy of most modern scientists.* The important thing for you to consider, if I may suggest it, is that for the mechanist, a *description* of

---

*See, for example, the physicist Richard Feynman in his famous book about quantum electrodynamics, *QED*: ". . . nearly all the vast apparent variety in

the world is identical to an *explanation* of the world."

"*For a mechanist, a description of the world is identical to an explanation of the world.* I haven't heard it put quite that way. That's very concise. And for the vitalist?"

"For a vitalist and teleologist who believes that life and its processes cannot be explained by the laws of physics and chemistry alone, there is need for explanation beyond description; there are those nagging questions, Why? and To what end? The division between these two views pervades the history of science down to our own time, and is one of the great themes of the dispute between science and religion. Naturally, it played an important part in the Galileo mess."

She shook her head in what seemed to be sadness at the very thought, then paused to take a delicate sip of coffee before carrying on.

"Aristotle was also what today we would call a 'scientific realist,' in that he believed that the world that we experience empirically, through our senses—and, of course, as science sees it through its instruments—actually exists, and is not a contrivance of our minds. This view he shares with modern science. And it is a view also shared by Galileo. The Church, on the other hand, adopted a Platonic or Neoplatonic view of this. The Church said that what we experience empirically is only a shadow or an image of a reality that exists behind it or beyond it. There were important reasons for this and it is, I believe, one of the more crucial distinctions in the history of philosophy and religion. But we can talk of that later, perhaps."

I nodded assent while she carried on.

"Incidentally, Aristotle's realism stops where the Earth's atmosphere ends. While his biology, for example, was in today's terms first-rate science, his cosmology cannot really be called science at all. His physical conception of the universe is influenced by the Pythagoreans and their notion of the circle and sphere as the most perfect figures. He regarded the heavens as a series of spinning spheres within spheres, each made of crystal, and each dependent on the others for its motion. Aristotle's universe was thus finite in space, in the sense that it is contained within an outer sphere. But

---

Nature results from the monotony of repeatedly changing just these three basic actions," the movement of photons, the movement of electrons, and the interaction between photons and electrons (Princeton University Press, 1986).

it was infinite, or at least indeterminate, in time—it had always existed and was not subject to destruction."

Her obvious enthusiasm for the subject matched my own. I was becoming very excited by what I was hearing. "Sister," I said, "I believe we're on exactly the same wavelength here. What got me interested in Galileo in the first place was a feeling that the mythology has missed the central point of the dispute with the Church. And you're absolutely right, it goes back to bedrock philosophical ideas. And your intuition about me is also correct. I don't know as much about Aristotle as I feel I should. Thank you for these notes. I'll read them with pleasure."

She seemed delighted. "You are very welcome, I'm sure. I really must fly now, but the important point I feel I ought to stress is that while Aristotle's science and in particular his teleology—the idea of purpose at work—was accepted by the Church after Aquinas, his scientific realism was not. The Church instead retained the age-old wisdom of St. Augustine, which was in turn based on Plato and the Neoplatonist philosophers. It was an essentially anti-realist point of view. This means that the Church did not accept that hypotheses that are based on observation and experiment could ever be the last word on reality. What science gives us are models that describe, usually in mathematical terms, what is observed in nature. But it does not tell us about what has been called 'the nature of nature.'"

With that, we said our goodbyes, agreeing to meet again in Venice. Sister Celeste rushed off in the direction of the railway station, and I strolled back toward my hotel, my head buzzing with caffeine and fresh ideas.

SEVEN

The Stagirite • The Primacy of Mind
The Good • Deductive Science

That afternoon at siesta time, I propped myself up on my bed and began reading the papers the little nun had given me. Marginal notes, amendments, additions, and revisions of all kinds abounded. Revised and re-revised and revised yet again, the manuscript showed unmistakable signs of that pitfall of academic authorship, the failure of nerve that comes with a deep understanding of any subject. "How," the writer asks, "can I possibly add anything of value in an area in which so many know so much and have written so well?" It was somewhat stilted but eminently readable, despite the notes and the numerous typographical errors, corrected in barely legible handwriting, and the fact that it appeared to be a second- or third-generation photocopy.

"Aristotle was born in 384 B.C. in the remote Greek settlement of Stagiria in Thrace," the manuscript began.

> We sometimes call him "the Stagirite" for that reason. His father had been court physician to the Macedonian warrior King Amyntas III—father of Philip of Macedonia and grandfather of Alexander the Great—a family connection that would be important in his later life. At eighteen years of age he was sent by his family to Athens to acquire a formal education, a practice

that had become popular with the rise of a new professional class of educators known as the Sophists. Legend has it that the young Aristotle squandered unfortunate amounts of time and money in the fleshpots of the city before finally enrolling himself in what was even then the most famous of all institutes of higher learning, the Academy of Plato. He became, by all accounts, an earnest student: according to Diogenes Laertus, when Plato read to the Academy his treatise on the soul, Aristotle "was the only person who sat it out, while the rest rose up and went away." He studied, and flourished, under the master for nearly twenty years until Plato's death in 347 B.C.

Aristotle was by then about thirty-five. He moved to the city of Atarneus in Asia Minor at the invitation of Hermias, a fellow alumnus of Plato's Academy and the city-state's self-made ruler. He married Hermias' daughter Pythias and set about establishing a school along the lines of the Academy. Work had barely begun when the city fell to the Persians, who captured and crucified Hermias, on suspicion of plotting to assist Philip of Macedonia's proposed invasion of Asia. Aristotle fled to the island of Lesbos, where in two productive years he seems to have done much of his remarkable work on natural history. His young wife, Pythias, died after giving birth to their daughter, and Aristotle later set up housekeeping with the courtesan Herpyllis, who bore him a son and with whom he would spend the rest of his life. We are unfortunately uncertain whether he was formally married. But he retained a tender devotion to Pythias and at the end asked that his bones be placed next to hers.

In 342 B.C. Philip of Macedonia invited Aristotle to the royal court his father had served, to tutor a precocious thirteen-year-old son who would become Alexander the Great. For three years the philosopher worked with the young man who was, like his father, an ardent admirer of Greek culture, and was soon to make his mark as one of history's great military geniuses, conqueror of most of the known world. We are told that Aristotle did his best to shape his headstrong pupil in the mold of the Homeric heroes like Ajax and Achilles, while enlightening him with the best of contemporary Greek thought.

In 335 B.C., when he was nearing fifty, Aristotle returned to Athens from Macedonia and established a rival school to Plato's Academy in the Lyceum, a gymnasium attached to the temple of Apollo Lyceus (god of shepherds), in a woodland on the southern outskirts of Athens. Classes were often held in the *peripatos* or covered walkway of the gymnasium, and the school is for this reason sometimes called the Peripatetic School. Aristotle himself is often referred to as "The Peripatetic" by association. The Lyceum became a center for research in every branch of knowledge, focusing especially on biology and history.

With the death of Alexander in 323 B.C. a wave of pent-up anti-Macedonian feeling swept through Greece. Aristotle, with his connections to the court, became a victim. Like Socrates before him he was charged with impiety or heresy. Unlike Socrates Aristotle did not remain in Athens to await condemnation and death. Declaring, it is said, that he wished to save the city from sinning against philosophy twice, he removed himself and some of his disciples to Khalkis, north of Athens, where he died a year later, aged sixty-two or sixty-three. His will has been preserved and in its generous provisions for his companion Herpyllis, his children, and his servants, not to mention his wish to be buried next to Pythias, it confirms the impression of a kindly man of genuine affection and strong emotion.

The scholars of the Middle Ages revered him as a thinker of unsurpassed insight, and his thought is incorporated into a thousand years of Christian theology and philosophy. Aristotle distanced himself from Plato's notion of the soul as an independent entity striving for release from temporary imprisonment in the body. He placed greater emphasis on the positive aspects of the material world and proposed that the soul is the source of life, which, when united with the physical body, forms the individual person. But not just humans had souls: for Aristotle vegetable souls determined such basic functions of life as nutrition, growth, and reproduction. Animal souls possessed these powers and added others of determining motion and feeling. Human souls added to all that the high privilege of reason. Vegetable and animal souls were transmitted from one generation to the next through reproduction, but the human soul, he said, came from "outside." This would be interpreted by Christian thinkers to mean it was specially created by God.

Aristotle saw divine or moral philosophy, as opposed to natural philosophy (or science), as the highest calling of the civilized man. First in the order of social evolution came the necessary attention to the provision of the necessities of life. Next came the development of the arts, and then politics. With the advent of the orderly and peaceful state came philosophy, which initially is preoccupied with the physical nature of things, as were the pre-Socratic Greek thinkers. Finally, philosophy lifts its gaze to a search for the nature of reality, the ultimate causes of things, and the purpose that underlies all change or motion.

But with the rise of science in the seventeenth and eighteenth centuries, academic history tried to claim him more or less exclusively for science. As a biologist and naturalist Aristotle was indeed astonishingly perceptive and decidedly modern, that is to say scientific, in his attitudes. His confession of ignorance as to the method of generation of bees (in *On the Generation of Animals*) is often quoted as an exemplary statement of the scientist's reliance on observation, or empiricism: "The facts have not yet been sufficiently established. If ever they are, then credit must be given to observation rather than to theories, and to theories only insofar as they are confirmed by the observed facts."

As a physicist and cosmologist, however, Aristotle has since Galileo's time been increasingly dismissed as quaintly absurd. Some modern analysts of his work have claimed that he himself renounced much of his religious thought in favor of an emergent empiricism, or scientific realism. A view that was widely held by the mid-twentieth century, this has in recent years been discredited as overly prejudiced by modern scientistic attitudes. Aristotle and his achievements are now more correctly seen against the backdrop of the thinkers who preceded him, rather than from the perspective of modern scientistic culture.

I put the manuscript aside for a moment to reflect on why Aristotle had always seemed a more modern philosopher than Plato, though their lives had overlapped in time. Of course his interest in science and his focus on observation were obvious clues. But there was another reason. While Plato had revered his mentor, Socrates, Aristotle had implicitly placed the pug-nosed founder of moral philosophy in a category with cranks and eccentrics, a "poor-spirited" man who had not had the gumption to claim the advantages he deserved, preferring to live a life of wretched self-denial. Aristotle's own "ideal man," on the other hand, had always struck me as a country-clubbish, self-righteous creature of decidedly material leanings. A bit of a stuffed shirt. Very modern, in fact.[1]

I continued to read:

> What distinguishes Aristotle from most modern scientific thinkers and practitioners is his insistence on the primacy of mind. For him, it is obvious that mind, as the comprehender of things, must come first and that no explanation of the nature of the world can succeed if it does not begin at this beginning. This is reflected in his conception of reality. Aristotle believed, like his teacher Plato, that all things are made up of matter and form. He had two very important ideas about the nature of matter, which find some resonance in modern quantum physics. The first of these ideas is that matter is an indeterminate substance that exists only as a potential (if at all) until form is imposed upon it, and it is thus "actualized." Because form is imposed by mind (i.e., it is an idea), any evidence of orderly processes in nature involving matter was in fact a reflection of the mind's innate structure as much as any preexisting order in external reality. In other words, reality as humans experience it was an interactive construct to which both potentialities in nature and the mind contribute. The second idea about matter is that it exists, not in static form, but always and only as a rhythmic coming into being and passing out of being, as it becomes realized in one *form* after another.* Mind is of course a very important part of this process, as well.
>
> This leads naturally to the notion of a conscious super-mind or deity operating behind the scenes. Though it may

---

*According to current theory, quanta have no existence until they are observed. Most have lifetimes of tiny fractions of a second and are constantly mutating.

sound unscientific and therefore nonsensical to the modern ear, this was a necessary element in Aristotle's science because he sought not just to describe the workings of nature, as modern science does, but to understand and explain them in terms of their ultimate purposes. This of course is a natural reflex, since it seems to be innate in human nature (and perhaps even a defining quality of the species) to want to go beyond the "how" of things to the "why." It is that striving, confined in our own time mainly to the arts and religion, that has produced many of the loftiest, most cherished achievements of civilization.

In this context, Aristotle's idea of the "good" as the ultimate goal of natural systems, of all motion, is easily understood in terms of an analogy used felicitously by the historian Giorgio de Santillana: "If we show a man a watch, he will not tend to explain it in terms of its machinery, unless he is a watchmaker or a very curious person. He will simply understand it as a timepiece. A watch is defined by its purpose, which is telling time, and that is also the 'good' of the watch. Our understanding [Aristotle said] tends thus to be formulated in terms of the 'good.'"[2]

The "good" of anything—a plant, a baby, a stone—is its achievement of its potential. An acorn seeks the good in becoming a mature oak tree. The good of an economy, we might say, is the provision of an acceptable level of comfort and security for all. The universe seeks ultimate good in terms of eternal order within its diversity.

This was of course different from the later Christian conception of the "good," which arose out of divine purpose. To the Christian, the good was that which helped to fulfill the divine intention. But the two ideas were by no means incompatible, requiring only the adaptation of Aristotle to the notion of an active rather than a passive God. Just as Platonic philosophy had done a thousand years earlier, Aristotle would give Christian philosophy a more rigorous philosophical undergirding.

Aristotle's physics is heavily influenced by his interest in biology. Or perhaps it would be more accurate to say that no strictly mechanical doctrine could have served his overall purpose, which was to present a comprehensive science of nature that included such modern disciplines as cosmology, physics, chemistry, and biology. In comparison, the mechanistic doctrines of Newton and his modern successors seem childishly simple in their strictly limited vision. When we refer to motion, for example, we normally mean a shift in the position of one object relative to another. For Aristotle, however, motion is a process of change that leads to the good. The essence of nature is motion in this sense of "change"—change whereby a thing becomes what it really is (growth), and change whereby a thing decomposes into material that becomes the basis for growth of something else (decay). His "motion" can of course also be motion in space, as when a stone falls or fire leaps up, but in each case the motion is part of the process of the thing's seeking and finding its proper place in the order of the universe. There is such a thing as "violent motion" as opposed to natural motion, as when a stone is ejected from a sling (or an arrow is released), but in all such cases natural motion eventually takes over and reestablishes equilibrium. If a thing is at rest, it means that it has reached its appointed place, its natural locus, or that it has been temporarily frustrated in its attempt to get there.

It is important to note that it was Aristotle's conception of motion in nature as always leading to the good that led him to reject the earlier mathematically oriented philosophies of thinkers like Pythagoras. It seemed to him that mathematics could not instruct us because it contained no mention of the good. Good is an end, implying motion, while mathematics deals with immobile entities, or stasis. There is no place in Aristotle for blind, mechanical determinism. His world presupposed mind, and a plan.

This paragraph was heavily underscored with a green felt-tipped marker. The phrase "mathematics could not instruct us because it contained no mention of the good" was circled as well as underlined.

In Aristotle's physics, all motion requires an exterior cause, a mover, without which movement comes to a stop. But his conception of cause is quite different from ours, because it is not mechanical. For Aristotle, "cause" is defined as "that without which something could not be." He uses the analogy of building a house, and asks, What are the causes of its construction? He answers that they are the building materials; the builder; the architect or designer; and finally the initial decision to build the house. (These reasons he names, in order: "material," "efficient," "formal," and "final"—we might also say *ultimate* —causes). These four kinds of causes, he said, are found throughout nature. This interpretation of causes of course implies the involvement of mind in nature, as in human activities like the building of a house.*

Aristotle's concept of motion carries within it two important consequences: first, there must be, ultimately, a prime mover that needs no outside assistance but is the ultimate cause of all motion—the "unmoved mover"; and second, the linkage of motion and mover ultimately connects up everything in the universe. Aristotle's universe is one vast, roistering, interdependent dance of generation, life, and decay, a cosmic striving for perfect order. It is a hierarchical universe, in which the upper orders constantly beam down motion and order as, for example, via the Sun's warmth and position in the sky, and yet are never diminished by their output nor affected by the orders beneath them. It is a fully achieved form that operates eternally under the auspices of the active intellect or consciousness of God, the ultimate good.

And here another very important technical point must be noted. In searching for a basic substrate or foundational material upon which to construct an understanding of the world, Aristotle rejected not only the Pythagorean idea of number as the basic building block but also the idea of indestructible, primal particles or "atoms" as proposed by Democritus and others. Instead he found it more useful to think of the natural order in terms of classifications and definitions. This is of course a completely different approach to understanding from modern scientific thought, but it has the advantage of being essentially unrestricted in its scope: it can encompass all of science,

---

* Once again, there is a resonance here with today's quantum physics, in which traditional ideas of cause and effect break down, and "effect" may precede "cause."

philosophy, and religion. Once again, I will use the excellent analogy provided by the historian de Santillana: "For instance, a man may be tall or short, musical or not; these are 'accidents.' He is certainly two-legged, but so is a hen; that is an 'attribute.' But he, and he alone, is an animal endowed with reason; that defines him, and in terms of more general concepts such as 'animal' and 'reason.' It is out of the particular definitions and their logical connections with the whole order of defined entities that Aristotle hopes to derive the natures and properties of things, very much as a lawyer might derive conclusions from the definition of legal entities."[3]

For Aristotle, the realm of the universe beyond the Earth was fundamentally different, because it seemed never to be touched by change and decay and generation. It remained constant (or so it seemed to him), and from this he concluded that the heavens consisted only of completely actualized forms. We might say "perfect" forms. He postulated a heavenly realm of transparent crystalline spheres in which the planets are transported on their paths through the skies. The stars are lodged in another, farther sphere, which is in turn enclosed and moved by the sphere of the outer heaven, also called the prime mover. Beyond this there is only the presence of God, whose fully actualized thought *is* the universe. There is no void and no space out there because the universe, as an achieved form, excludes the presence of anything beyond it.

Aristotle's God is not a Creator because His universe has no beginning—cannot have one because that would imply movement or action on the part of what must be an unmoved mover. Nor is God Providence in our Christian sense, because He is beyond and above nature and His life is not one of action and intervention but of contemplation. In fact [as the historian W. G. de Burgh writes], "he does not even know the universe but is pure self-conscious intelligence at once the subject and the object of his own most perfect thought. Alone in this absolute transcendence God draws the world toward himself as the goal of its desire. 'The final cause produces motion as does an object of love [Aristotle says], and through that which it moves [i.e., the outer sphere of the stars], moves all things else.'"[4]

For Aristotle the physicist, certain aspects of the physical universe as modeled by some of his mathematically minded predecessors such as Pythagoras and Anaximander seemed preposterous. What we now call the law of inertia, the idea that a thing continues to move of its own accord as long as nothing interferes with that motion, was nonsensical because it meant that there could be motion that had no goal and no cause. Nor could the universe be infinite in scope, because in an infinite universe there is no up or down, all directions are the same, and things would not be able to find their way to their natural locus. Without that ability there could be no order, only perfectly dispersed uniformity, which is a form of chaos. But, clearly, there *was* order in the universe—it was visible on every hand both on Earth and in the heavens. In fact, the word *cosmos* meant order and harmony.

As we have seen, Aristotle did not put much stock in mathematics, and the idea of modeling nature in mathematical formulas in order to study it more deeply seemed at best marginally useful. When a modern scientist watches a ball roll down an inclined plane, he sees in his mind what he can never see in reality—frictionless motion, which is a mathematical abstraction. The laws he postulates to explain his observations apply strictly only in his imaginary realm of mathematical models, and only approximate reality as we experience it.* Aristotle, observing the same experiment, saw an object seeking its proper locus, that is, its place at the center of the universe, which was the center of the Earth. Why construct arbitrary, abstract archetypes when the only model needed was there in plain sight for everyone to see—nature as we experience it every day? This sensible reality was Aristotle's starting point, the given, and from here he stretched his mind and imagination in trying to understand the meaning of the wonders he observed.

*The physicist and historian of science Alexander Koyré expresses this view, which is shared by Plato, in his book *Metaphysics and Measurement*: "It is impossible in practice to produce a plane surface which is truly a plane; or to make a spherical surface which is so in reality. Perfectly rigid bodies do not, and cannot, exist in *rerum natura*; nor can perfectly elastic bodies; and it is not possible to make an absolutely correct measurement. Perfection is not of this world; no doubt we can approach it, but we cannot attain it. Between empirical fact and theoretical concept there remains, and will always remain, a gap that cannot be bridged."

It was not a science that encouraged discovery in the sense in which *we* understand it but rather a deep study of what was already known. The undeveloped was to be explained in terms of the fully developed, and not, as in modern science, the other way around. The plant, the achieved form, is prior to the seed, as the man is prior to the child. Aristotle's science reasoned deductively, from the top down, from the complete to the incomplete, as opposed to modern science's inductive or bottom-up approach, in which phenomena are reduced into their most basic elements, each of which is studied separately.

Aristotle's argument for God's existence from the observed fact of motion was to become the foundation of Christian theology in the Middle Ages. His exact language is taken up by St. Thomas Aquinas, the great theologian of the thirteenth century (in *Summa contra Gentiles*), and the poet Dante (in *Paradiso*) when each refers to "the Love that moves the sun and other stars." But the adoption of Aristotelianism was fraught with no less conflict and anxiety than was the introduction of Copernican astronomy and Galileo's scientific realism four centuries later.

At that point in the text fatigue must have overcome the excitement I was feeling. The next thing I was aware of was the sound of activity in the street beyond my window, as shop-owners reopened their doors and customers filled their shopping bags with meat, fish, produce, wine, and the other necessities for that night's dinner.

EIGHT

Aristotle's Influence • A Question of Infinity
Copernican Caution • Saving the Appearances

By the sixteenth century, the hybridized Aristotle in the theology of Aquinas had become bedrock dogma of the Church of Rome. (Aquinas was made a Doctor of the Church in 1567.)

There was much of Aristotle in Copernicus. Enough, in fact, to mollify the Church of Rome, anxious though it was in those times of Protestant schism to root out ideas that might undermine its authority. Copernicus' ideas did appear scandalously heterodox on the surface. But looked at from an Aristotelian point of view, the displacement and setting in motion of the Earth might be considered simply a mathematical convenience, a useful hypothesis that enabled mathematicians, astronomers, and others concerned with the movements in the heavens to improve the accuracy of their calculations. That, initially, was the position adopted by the Church. Copernicus, far from being persecuted, was invited to give his opinion on calendar reform to the Lateran Council of Church authorities in 1514, and Pope Clement VII, in a letter dated 1536, formally urged him to publish his ideas on the workings of the heavens. Copernicus had been at work on his opus *De Revolutionibus* for years. Nevertheless, he demurred, to the intense frustration of his friends and admirers.

What was the cause of his reticence? What harm could there be in an alternative model of planetary mechanics? There were at least two problems that Copernicus must have been aware of. A careful survey of his text would reveal that although the system was presented in the form of a mathematical hypothesis like Ptolemy's, Copernicus himself had come to believe that it represented physical reality. He knew that this was a dangerous, possibly heretical position to take. This, however, was a minor problem, one that could easily be cleared up by the Church censors with the deletion of a few words.

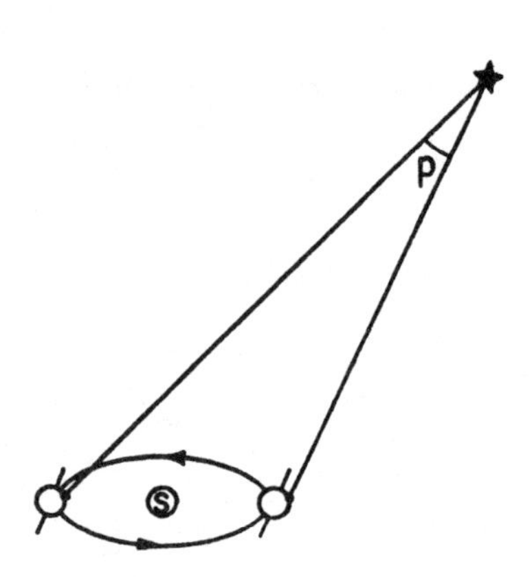

Stellar parallax. Because the line between an observer on Earth and the star moves as the Earth orbits the Sun, the star's apparent position should shift by the angle P over the course of six months.

There was another worm in the apple, and it had to do with something called stellar parallax. If the Earth was not fixed at the center of the universe and was in motion in a vast orbit around the Sun, there ought to be a perceptible shift in the observed positions of the so-called fixed stars through the seasons as the Earth moved from one side of its orbit to the other. Astronomers, try as they might, were unable to detect any such movement. Two conclusions could follow: either Copernicus was wrong and the Earth was in fact at the center of the universe and did not move, or the stars were so incredibly distant that from their perspective the path described by the Earth's orbit was a mere point, making the stellar parallax too tiny to be observable.* The second conclusion would (and did) naturally lead to speculation about an infinite universe. Copernicus could not avoid the accusation of promoting the idea.

The question of the infinite or finite nature of the universe was one of serious concern for the Church. A finite universe, no matter how constructed, can be conceived of as having been created, but an infinite universe cannot. It goes without saying that Scripture committed the Church to the notion that God had created the universe at a given time. An infinite universe was therefore worse than

* Stellar parallax was finally observed reliably in 1838 by the German astronomer Friedrich Wilhelm Bessel.

a contradiction, it was an abomination. It was a denial of God's creative action and, therefore, of the very existence of the Christian God.

Not long after Copernicus' death, Giordano Bruno would be burned as a heretic—tongue-tied and lashed to a stake in Piazza Campo dei Fiore in Rome. It was—in part—for stubbornly espousing an infinite universe that the renegade Italian monk had run afoul of the Church authorities. In Bruno, the Church seems to have seen clearly and perhaps for the first time the terrible fruit of the Copernican hypothesis. If the universe were not finite, where would Heaven and Purgatory be located? How could God be separate and distinct from His creation, as Christian theology insisted He was? Since an infinite universe would contain everything there was, God would have to be a part of it. If He was part of it, how could He have created it? Furthermore, order and harmony simply could not be found in an infinite, and therefore completely formless and uniform, universe. The entire Aristotelian-Ptolemaic-Thomist cosmology would collapse and with it the thousand-year-old hierarchy of interdependent existence linking in an unbroken chain the inanimate to plant and animal life, to man, to angels, to God. And if that happened, what would become of the painstakingly crafted system of earthly values and morals built on those assumptions? The implications were clear, and not just to Church scholars. Shakespeare wrote in *All's Well That Ends Well* (c. 1602), "They say morals are past; and we have our philosophical [i.e., scientific] persons, to make modern and familiar, things supernatural and causeless."

A modern reader may empathize with this distraught reaction to the idea of infinity, if only for having seen its mirror image among scientists of the twentieth century. By that time the once-novel seventeenth-century hypothesis of a steady-state universe infinite in time and space had become solidly entrenched. By the middle of the twentieth century, however, three separate lines of inquiry into the motions of the galaxies, the laws of thermodynamics, and the life history of stars came together to point to one conclusion: the universe had a beginning and is therefore finite. And the best answer to the question, What came before the beginning—the famous "Big Bang"?—was, nothing. Astronomers were confronted with the likelihood that the universe was created ex nihilo, out of nothing. That of course raised the possibility, some would say the unavoidable necessity, of a Creator.

Like their churchly predecessors, astronomers—indeed, scientists in general—found the paradigm shift that was being forced on them to be extremely vexing, and their published reactions are laden with bewilderment and emotion. Writing in the 1970s the astronomer Robert Jastrow notes that, "when an astronomer talks about God, his colleagues assume he is either over the hill or going bonkers." He points out that Albert Einstein found the idea of an expanding and therefore finite universe "annoying," though he was forced to accept it late in his life. (Like most of his contemporaries in science, Einstein subscribed to a theistic notion of God akin to Descartes' prime mover or clock winder. Only within such a cosmology could religion and modern science peacefully coexist, since the tenuousness of their contact made it possible for them to ignore each other. When in 1921 the archbishop of Canterbury asked him what he thought were the implications of his Theory of Relativity for theology, he replied, "None. Relativity is a purely scientific matter and has nothing to do with religion.")[1]

The British astronomer Arthur Eddington said of the Big Bang, "I have no axe to grind in this discussion . . . [but] the notion of a beginning is repugnant to me. . . . I simply do not believe that the present order of things started off with a bang . . . the expanding universe is preposterous . . . incredible . . . it leaves me cold." The German chemist Walter Nernst wrote, "To deny the infinite duration of time would be to betray the very foundations of science." Jastrow reports other representative reactions: Phillip Morrison of MIT said, "I find it hard to accept the Big Bang theory; I would like to reject it." Allan Sandage of the Palomar Observatory, whose own observations had helped establish the theory of an expanding universe, said, "It's such a strange conclusion . . . it cannot really be true."[2]

Because his hypothesis implied the possibility of an infinite universe, Copernicus, then, had good reason to be worried about the potential reaction to his book. It was only in 1543, when he saw that his life was nearing an end, that he finally succumbed to pressure from friends and colleagues and agreed to have his manuscript published. He put it in the hands of a loyal disciple, Georg Joachim Rheticus, who took it to Nuremberg for printing. Rheticus entrusted the printing to the leading theologian of Nuremberg and a fellow Lutheran, pastor Andres Osiander. Osiander, an admirer of Copernicus, took it upon himself to write an anonymous preface to the book, which has the form of a disclaimer: "For these hypotheses need not be true or even probable—it is sufficient that they should save the appearances. Nevertheless these hypotheses deserve to be known along with similarly speculative theories of the ancients, because they are admirable and also simple, and bring with them a huge treasure of very skillful observations." Osiander concludes, "So far as hypotheses are concerned, let no one expect anything certain from astronomy, which cannot furnish it, lest he accept as the truth ideas conceived for another purpose [i.e., mathematical convenience], and depart from this study a greater fool than when he entered it."

It is a part of the mythology of scientism that the death of poor, timid Copernicus was hastened by reading this preface. (A copy of the newly printed book is said to have been brought to the astronomer on the last day of his life, May 24, 1543.) Its appearance in the first editions of *De Revolutionibus* has been called the greatest scandal in the history of science. The suggestion is that aged and ailing Copernicus had no advance knowledge of it and would never have approved its inclusion in his book. However, there is reliable evidence that Copernicus did read the preface prior to publication. On the other hand, there is no evidence that after reading it, he objected or requested changes. Beyond that, there is the fact that he himself cautiously prefaced his work "For Mathematicians Only." Surviving correspondence shows that Osiander was stating in the preface what he had discussed with Copernicus in person and in letters—the classic formula for the coexistence of science with religion. And that is that science provides not truth but useful mathematical fictions that enable people to better understand and manipulate the world around them; truth is the purview of philosophy and religion. If Copernicus disagreed with this formulation, he has left us no evidence of that fact.[3]

NINE

The Padua Years • The Telescopic Discoveries The Lure of the Medici • The Jesuits Rumors of Reaction

As his teaching contract was ending at the University of Pisa, Galileo found himself, again, in serious financial straits. The death of his father in 1591 made him, as eldest son, responsible for his mother, brothers, and sisters. In that same year his sister Virginia was married, and it was Galileo who had to provide her dowry "in measure beyond his means." The obligation consumed most of his salary for the next several years. Later, his second sister, Livia, would marry, and once again a huge dowry was expected. Galileo asked for a two-year advance on his professorial salary and signed a hefty nuptial contract jointly with his brother Michelangelo, a feckless if talented musician. But Michelangelo not only failed to provide his part of the money but was himself frequently in need of help from his older brother to support his wife and children.

In 1592 Galileo applied for and won the post of chair of mathematics at the University of Padua, at 180 florins a year, later raised to 520. (The leading philosopher at the university was paid 2,000, a disparity that gives some indication of the continuing low status of mathematics in academia.) The university was a center of learning renowned throughout Europe for its devotion to Aristotelian teachings in science and medicine. He remained there for eighteen years in the liberal environment of rule by the republic of Venice, teaching geometry, astronomy, and military engineering and continuing his researches into the laws of motion. He published papers on mechanics, fortifications, and spherical geometry, and their brilliance earned him a spreading fame. He found himself more and more frequently in Venice's Arsenale, providing engineering advice to the craftsmen who toiled in the shipyard.

His domestic life in Padua could only have been chaotic. To supplement his teaching and consulting income, he tutored the sons of aristocrats from all over Europe, often taking them into his own home as boarders—fifteen or twenty at a time. Among his students was Cosimo de' Medici, the young son of Grand Duke Ferdinand, ruler of Florence, a student he had actively sought out with a letter to the grand duke. Galileo had invented a very handy calculating device akin to a slide rule, which he called a geometric compass, and which the English called a "sector." These he manufactured by the hundreds in a workshop he built in his courtyard. To perform the casting and engraving he hired a metalsmith from the Venice Arsenale who arrived in Padua with his wife, children, and servant in tow, all of whom joined the already boisterous ménage in Galileo's home and remained there for the next eleven years. It is little wonder that Galileo chafed at not having enough time to pursue his mathematical and mechanical research. Nevertheless, looking back from the perspective of old age, he would describe his Padua years as the happiest of his life.

It was during this time that he met Marina Gambina, who became his mistress and who bore him three children. The first, Virginia, was born August 13, 1600, when Marina was twenty-two and Galileo was thirty-six. A year later, Antonia was born, and five years after that there was a son, Vincenzo. Galileo acknowledged paternity of the children, though he never married and in fact never lived in the same house with Marina. At first she lived in Venice and he rode up to visit on weekends. Later she moved into a house that he provided for her just a short walk from his own in Padua. When he moved to Florence to take up a new position in 1610, he ended their relationship. Galileo placed his two daughters in a cloistered convent near Florence, which they were forbidden to leave during their lifetimes.* The girls were twelve and thirteen at the time, far too young for the Church to accept under normal circumstances: rules and tradition both dictated that girls should be old enough to decide of their own free will to enter a convent. But Galileo had by then achieved enough social standing to be in a position to pull the necessary strings within the ecclesiastical hierarchy. Virginia, who took the name of Sister Maria Celeste, adjusted to the regime of seclusion and poverty as well as might be expected; her younger sister appears to have suffered psychologically and was prone to frequent bouts of mental and physical illness. There exists an extensive correspondence from Sister Maria Celeste to her father, but Galileo's letters of reply have been lost. What has survived depicts a tender and caring relationship. This seems to have intensified in Galileo's later life, and he visited her frequently.† For his son, Vincenzo, he performed the expected paternal duties of financing an education and helping out with the purchase of a first home after his marriage, but the two do not appear to have been close.

Galileo did most of his breakthrough experimental work on mechanics at the University of Padua, intending always to publish it soon. But in 1609, when he was forty-five years old, fate intervened and delayed the book for nearly thirty years. His encounter with a Dutch optical toy that people were calling a "spyglass" in that year swept him into a vortex that made him, overnight, Europe's most celebrated scientist and sharply altered the trajectory of his career.

By this time the grand synthesis of Copernicus had been around for more than half a century and seemed a dead issue. It had been widely read, but not because of the Sun-centered hypothesis so much as in spite of it. The reaction of the English astronomer

---

* "He Himself deigned and willed to be placed in a sepulcher of stone. And it pleased Him to be so entombed for forty hours. So, my dear Sisters, you follow Him. For after obedience, poverty, and pure chastity, you have holy enclosure to hold on to, enclosure in which you can live for forty years either more or less, and in which you will die. You are, therefore, already now in your sepulcher of stone, that is, your vowed enclosure." *The Testament of St. Colette.*

† In her book *Galileo's Daughter*, which focuses on the correspondence, Dava Sobel provides an account of this relationship (Walker, 1999).

Thomas Blundeville was typical: "Copernicus . . . affirmeth that the earth turneth about and that the sun standeth still in the midst of the heavens, by help of which false supposition he hath made truer demonstrations of the motions and revolutions of the celestial spheres than ever were made before."[1] Tycho Brahe, Europe's acknowledged king of astronomy, had declared himself opposed to the hypothesis and proposed a system of his own in which the Earth remained at the center of things. Most academics dismissed the Copernican model because it could not be squared with Aristotelian physics. The Protestant Churches denounced it because it seemed to contradict certain passages in Scripture, literally interpreted. The Roman Catholic hierarchy's attitude to Copernicanism was more complicated but could be summed up as benign indifference. Though Copernicus was respected as a mathematician and a churchman, his planetary system was regarded as one more mathematical fiction that might be helpful in such tasks as calendar making but could hold no serious claim to being a description of physical reality.

In his astronomy classes at Padua Galileo was obliged to teach the Ptolemaic system just as he had been at Pisa, but by now he had become convinced that the Copernican model was superior both in a purely mathematical sense and as a description of how the world really was. We know this because he said as much in a letter to the like-minded Kepler in 1597:

> I have been for many years now a follower of the Copernican theory [since it explains] the reasons of many phenomena which are quite incomprehensible according to the views commonly accepted. I have written up many reasons and refutations on the subject, but I have not dared until now to bring them into the open, being warned by the fortunes of Copernicus himself, our master, who procured for himself immortal fame among a few but stepped down among the great crowd (for this is how foolish people are numbered), only to be derided and dishonored. I would dare publish my thoughts if there were many like you; but, since there are not, I shall forbear.[2]

Galileo's concern was that he would be ridiculed by the academic establishment. (Hindsight might suggest he feared persecution by the Church of Rome, but he had no basis for such a fear at the time of writing.) Kepler's moving reply must have given him pause:

> I could only have wished [Kepler wrote] that you, who have so profound an insight, would choose another way. You advise us, by your personal example, and in discreetly veiled fashion, to retreat before the general ignorance and not to expose ourselves or heedlessly to oppose the violent attacks of the mob of scholars (and in this you follow Plato and the Pythagoreans, our true preceptors). But after a tremendous task has been begun in our time, first by Copernicus and then by many very learned mathematicians, and when the assertion that the Earth moves can no longer be considered something new, would it not be much better to pull the wagon to its goal by our joint efforts, now that we have got it underway, and gradually, with powerful voices, to shout down the common herd, which really does not weigh the arguments very carefully? Thus perhaps by cleverness we may bring it to a knowledge of the truth. With your arguments you would at the same time help your comrades who endure so many unjust judgments, for they would obtain either comfort from your agreement or protection from your influential position. It is not only your Italians who cannot believe that they move if they do not feel it, but we in Germany also do not by any means endear ourselves with this idea. Yet there are ways by which we protect ourselves against these difficulties. . . .
>
> Be of good cheer, Galileo, and come out publicly. If I judge correctly, there are only a few of the distinguished mathematicians of Europe who would part company with us, so great is the power of truth. If Italy seems a less favorable place for your publication, and if you look for difficulties there, perhaps Germany will allow us this freedom. . . .[3]

More than a decade would pass from the time of that exchange between the two scientists and Galileo's introduction to the telescope. During those years Galileo maintained a discreet silence. But then fate presented him with discoveries of such magnitude that he had no choice but to make them public. Indeed, his fear seems to have become that someone else would make the same discoveries and publish them before he did.

It has been sensibly suggested that the invention of the printing press in the fifteenth century made the invention of the telescope inevitable. The reasoning goes like this: the increased availability of books inflated the demand for eyeglasses to the point that there were many opticians all over Europe who were experienced at grinding lenses of various kinds. The effects of placing one lens in front of another and varying the focal length are so striking that it is impossible to believe that the principle of magnification of distant objects had not been noticed on many occasions. Prior to this theory coming into vogue, it used to be said that the telescope was invented by Hans Lippershey in the Netherlands in 1608. But this is apparently not the case; his application to the Estates General of Holland for a patent in that year was turned down on grounds that "many other persons had a knowledge of the invention." In that year, a telescope with a magnifying power of 7x was offered for sale in Frankfurt. In 1609 small telescopes were being sold as novelties in Paris, London, and several cities in Germany.*

It was during a visit to Venice in July of 1609 that Galileo was told of the device. He had been in the city to try to negotiate a salary increase. Exactly how Galileo heard about the telescope, or from whom, is not known, but immediately on his return to Padua on August 3, he constructed one of his own with a 3x magnifying power and quickly improved on this with another of 8x. By August 21 he was back in Venice to invite the Senate of the republic to a demonstration of a telescope of his own making on the tower of St. Mark's. The presentation was a huge success. Senators were astonished to find that they could clearly see the cupolas, bell tower, and classical facade of the church of San Giustina in Padua, twenty-two miles away. Turning the device to the sea, they were able to see the sails of incoming ships two hours or more before they became visible to the naked eye, and the military importance of this was not

---

*Galileo used the word *occhiale* for his telescope; an approximate English translation might be "spyglass." In Latin this became *perspicillum* or *arundo optica*. The Greek name *telescope* was proposed several years later by a member of the Lincian Academy in Rome.

lost on them. Three days later Galileo gave his telescope to the city, along with a letter outlining how it worked. A grateful Senate, in an outburst of public largesse, doubled Galileo's professorial salary to 1,000 florins a year. It was only after the increase had been promised that the politicians learned that telescopes had been available in several European centers for some time. They were not pleased. When Galileo received his revised teaching contract he found that while the promised raise was there, it would not take effect for another year, and it was formally stipulated that this was the last raise he would ever get.

The humor in the situation was understandably lost on Galileo, and no doubt the Senate's churlishness contributed to his eventual decision to leave the University of Padua. For the moment, however, he succumbed to the fascination of the new instrument. He soon was producing models with 30x magnifying power. It is unclear why, or even whether, Galileo was the first to turn the telescope to the skies. Some historians suggest that his skill as an inventor and craftsman allowed him to produce superior lenses, which provided the resolution necessary to see astronomical objects clearly. There is a certain plausibility in this: there was a side of him that prefigured such protean inventors as Edison and Marconi, and one thinks of him more as a tinkering engineer than an academic researcher. There is also good evidence that he never thoroughly grasped the principles behind the device. The unassuming Kepler, who did understand the theory of optics, was able to greatly improve on Galileo's design, and it was the Keplerian model that was adopted for general use by astronomers during Galileo's lifetime.

Nevertheless, Galileo was first to capitalize on the device's potential. In late 1609 and early 1610 Galileo excitedly announced a bewildering series of discoveries he had made while observing the night sky. He found that the Moon's surface was irregular, populated by mountain ranges and craters. It was, apparently, much like Earth, an observation that called into question the age-old Aristotelian certainty that the Earth and the celestial bodies were made of fundamentally different materials. He trained his glass on the pearly streak called the Milky Way—for medieval pilgrims a ribbon of light showing the way across the Pyrenees to the shrine of

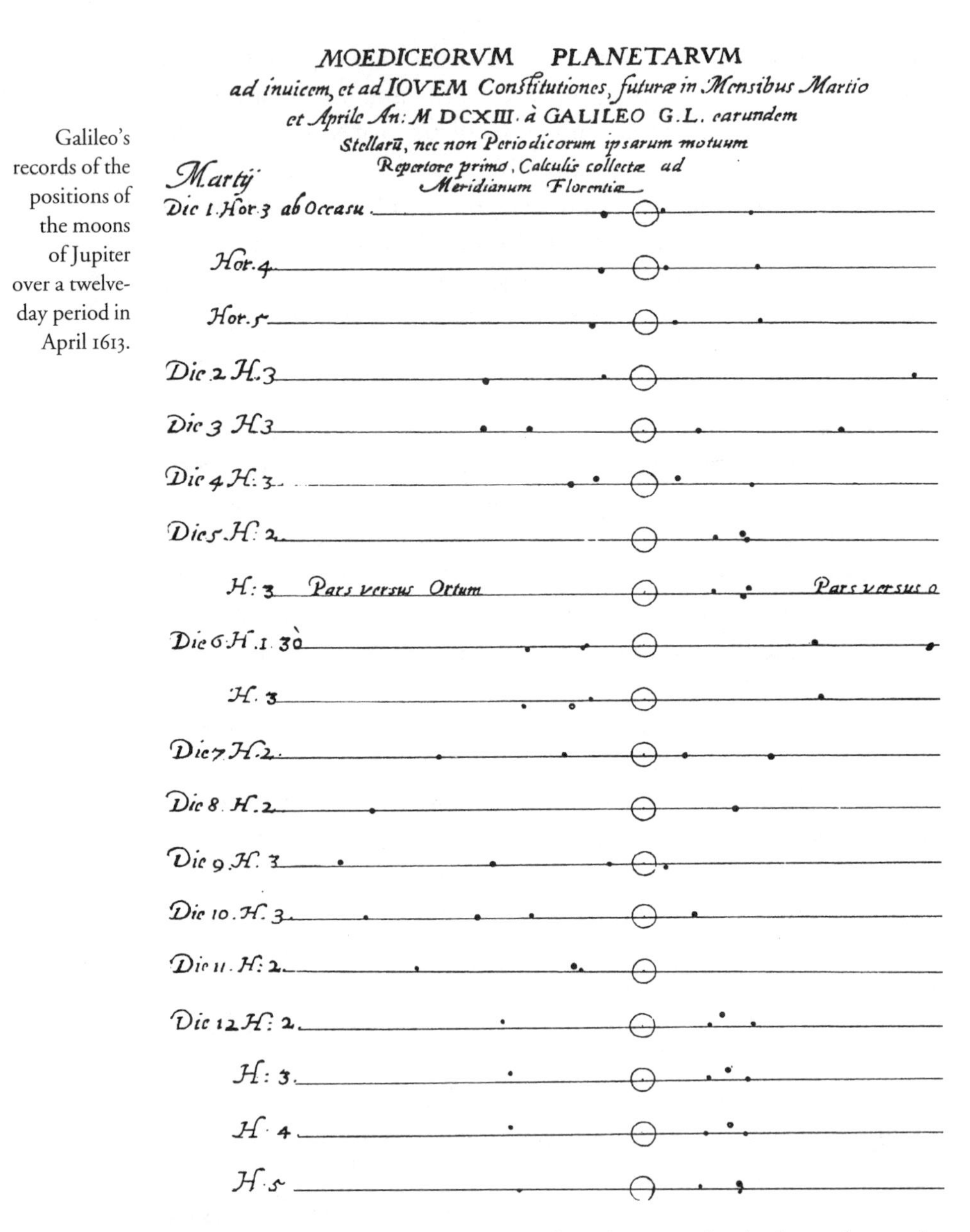

Galileo's records of the positions of the moons of Jupiter over a twelve-day period in April 1613.

Santiago de Compostela, and for the Greeks before them the pathway to the palace of the gods—and he discovered it was composed of myriad of unimaginably distant individual stars. He found that there were satellites orbiting Jupiter, as the Moon orbits the Earth.

If Galileo had harbored nagging reservations about the Copernican system in light of the many objections raised by Aristotelian academics and theologians—and the evidence would suggest he did—what he saw through his telescope changed his mind. It wiped out almost all the astronomical objections, all except the problem of stellar parallax, which could anyway be accounted for in an infinite universe.

The Copernican system implied that Earth must be like the other planets, because it was no longer the center of the universe but precisely just another planet. Galileo's observations of the imperfections of the Moon (and, later, the spots on the Sun) lent powerful support to the idea that the heavenly bodies were not the perfect, immutable entities Aristotle had surmised but were all essentially of the same stuff as Earth.

Another persistent objection had been that the behavior of Mars did not correspond with what would be predicted by the Copernican system. As Earth's outer neighbor, Mars should have been relatively closer to Earth when both planets were on the same side of the Sun, and more distant when on the opposite sides of the Sun. The difference in proximity was great, and so there ought to have been a marked difference in the size of Mars as seen from Earth. But only a small difference was visible to the naked eye. In the telescope, however, Galileo could see that the variation was every bit as dramatic as Copernicus had predicted it would be.

Each of these discoveries tended to undermine the Aristotelian-Ptolemaic cosmology, but most damaging of all was the detection of Jupiter's moons. The Copernican system arranged the planets neatly around the Sun according to the speed of their revolutions —Mercury fastest and closest; Saturn slowest and farthest—and here was Jupiter, a miniature solar system with satellites arranged sequentially by period in exactly the same way. It was powerful circumstantial evidence. As well, one of the more important arguments against the idea of the Earth moving around the Sun was that if it did, it would quickly leave behind its Moon. The idea of the Moon orbiting the Earth while the Earth traveled at great speed around the Sun seemed completely implausible to anyone versed in Aristotelian physics. But here was Jupiter, which everyone agreed revolved around something—either the Sun or the Earth—and its satellites stuck with it. As Galileo noted:

> Here [in the Jovian moons] we have a powerful and elegant argument to quiet the doubts of those who, while accepting without difficulty that the planets revolve around the sun in the Copernican system, are so disturbed to have the moon alone revolve around the earth while accompanying it in an annual revolution about the sun, that they believe that this structure of the universe should be rejected as impossible. But now we have not just one planet revolving around another; our eyes show us four stars that wander around Jupiter as does the moon around the earth, and that all together they trace out a grand revolution about the sun in the space of twelve years.[4]

The quotation is from the publication he rushed into print in March 1610—a little book he called *Siderius Nuncius*, variously translated as *The Starry Messenger* and *Messenger of the Stars*. He was in his forty-seventh year, and it was his first major scientific publication. It was just a year after Kepler's difficult Latin tome containing the revelation of elliptical orbits, *Astronomia Nova*, had appeared and almost immediately dropped from sight, and the contrast between the two works could scarcely have been greater. Galileo's book, hardly more than a pamphlet, was composed in a clear and compelling prose style that would distinguish its author as one of his generation's finest writers. It was printed in an edition of 550 copies and sold out almost immediately.

It is hard to imagine the emotions Galileo must have felt as night after night his telescope revealed to him objects and images never before seen or suspected by even the greatest astronomers. Not only were these new phenomena amazing in themselves, but they strongly suggested to a knowledgeable observer like Galileo that the cosmology that had dominated Western thought for a thousand years was demonstrably wrong. Some small sense of his excitement can be gleaned from the passage in *The Starry Messenger* in which he announces to the world his initial observations of the Milky Way:

> With the aid of the telescope this has been scrutinized so directly and with such ocular certainty that all the disputes

> which have vexed philosophers through so many ages have been resolved, and we are at last freed from wordy debates about it. The galaxy is, in fact, nothing but a congeries of innumerable stars grouped together in clusters. Upon whatever part of it the telescope is directed, a vast crowd of stars is immediately presented to view. Many of them are rather large and quite bright, while the number of smaller ones is quite beyond calculation.
>
> But it is not only in the Milky Way that whitish clouds are seen; several patches of similar aspect shine with faint light here and there throughout the ether, and if the telescope is turned upon any of these it confronts us with a tight mass of stars. And what is even more remarkable, the stars which have been called "nebulous" by every astronomer up to this time turn out to be groups of very small stars arranged in a wonderful manner. Although each star separately escapes our sight on account of its smallness or the immense distance from us, the mingling of their rays gives rise to that gleam which was formerly believed to be some denser part of the ether that was capable of reflecting rays from stars or from the sun. . . .[5]

Much has been made of the hostile reaction to Galileo's discoveries among conservative clergy and academics. However, for most of this we have only Galileo's testimony in his correspondence and in his later work, *Dialogue on the Two Chief World Systems*, and the embroidery of his earliest hagiographers. Recent scholarship suggests that his "opponents" were frequently straw men concocted as part of Galileo's polemical style.[6] The story has often been told of how philosophers in Pisa refused to look through the telescope when offered the chance, how others reluctantly looked and professed to see nothing. "My dear Kepler," Galileo wrote, "what would you say of the learned here, who, replete with the pertinacity of the asp, have steadfastly refused to cast a glance through the telescope? What shall we make of all this? Shall we laugh, or shall we cry?" One such philosopher (and there were only a handful) was Giulio Libri, who taught with Galileo at both Pisa and Padua.[7] When he succumbed to the infirmities of old age just a few months after turning down an opportunity to see the night sky through a telescope,

Galileo's uncharitable if amusing comment was that he hoped that Libri, having refused to look at the heavenly bodies while on earth, would take the opportunity while on his way to heaven.[8]

It is worth pointing out that the issue of whether to accept as real what was presented to the eye by the telescope was not quite as simple as it seems to the contemporary mind, accustomed as it is to all kinds of instrument-mediated observations (for example, television). As is often the case, the glib modern scorn of older ideas is based not on superior knowledge but ignorance of the subtleties of those ideas. It is one thing to accept the magnifying power of the telescope when it is trained on objects on Earth—objects that are known to exist. It is quite another to believe in what the telescope shows when it is pointed to the sky. For to do so means to believe in the existence of something that cannot be seen with the naked eye. The existence of previously unseen celestial objects and features, in other words, cannot be verified except with another, similar instrument. And why should *that* instrument be believed? A genuine leap of faith was involved, and it is not surprising that some found that leap both difficult and frightening. A deep epistemological issue had been raised: did the knowledge gained "instrumentally," as we would say—via the telescope—have the same status as knowledge based on unmediated sensory experience? Could we ever be as certain of the one as we were of the other? These were questions worth asking, then as now, though they are nowadays seldom discussed. (We know, for example, that the reality presented on television is in important ways different from the reality of firsthand observation.) Galileo had an answer, in fact the only possible answer for his time. And that was that the security of such knowledge was ensured in the process of repeated duplication of the initial experience. Logically, of course, this makes no sense. Simply repeating what may initially have been an erroneous procedure does not magically make it sound. It is a modern, instrumentally useful but logically and philosophically fragile answer to a profound question that continues to dog science into the twenty-first century. It is, in fact, not an answer at all in any philosophical sense—merely a rule of thumb. With this in mind it seems quite natural that there was initial resistance to Galileo's telescopic discoveries among clerics and lay philosophers steeped in the complexities of two thousand years of inherited wisdom. Only when observations had been confirmed by many different instruments in the hands of many different men in different locations was some measure of comfort established with Galileo's sightings. But for many, a nagging unease would remain.

Nevertheless, and perhaps surprisingly, a more typical response than the aged Libri's was the reaction from lay academics at Galileo's University of Padua, as he describes it in a letter to a friend: "As I mentioned to you in my last letter, I have given three public lectures on the subject of the [moons of Jupiter] and my other observations. The whole university turned out, and I so convinced and satisfied everyone that in the end those very leaders who at first were my sharpest critics and the most stubborn opponents of the things I had written, seeing their case to be desperate and in fact lost, publicly stated that they are not only persuaded but are ready to defend and support my doctrines against any philosopher who dares to attack them."[9]

The doughty Kepler had by now inherited Tycho's position as Europe's leading astronomer, and he immediately threw the full weight of his authority behind *The Starry Messenger* and its author. With voluble enthusiasm, he accepted on trust Galileo's descriptions of what he had seen.* And the response of the educated public could only be described as rapturously favorable. One historian recounts a story of a package being delivered to a scientific friend of Galileo's in Florence shortly after publication of *The Starry Messenger*. Neighbors and passersby, thinking the parcel might contain a telescope, badgered the recipient into opening it on the spot. When it turned out instead to be a copy of the new book, they insisted that he read it to them that very night, in his home.[10] Galileo's printers in Padua invited poetic tributes and were inundated; England's John Donne referred to the discoveries of "Galileo the Florentine" in his poem "Ignatius," published just ten months after *The Starry Messenger*:

Man has weav'd out a net, and this net throwne
Upon the Heavens, and now they are his owne . . .

---

*Even though Galileo continued to ignore Kepler's book, *The New Astronomy*, in the face of Kepler's repeated requests for feedback.

Editions of *The Starry Messenger* were printed and offered to eager buyers all over Europe, and by 1615 it was even available in China, thanks to Jesuit missionaries. An English physician, Nathaniel Highmore, commented: "The invention of Galileus was the wonder of the witts of the world and discovered heaven as if it had been under our feete. It made a fragment of glass of more value than the richest jewell."[11]

Such animosity as the book did arouse can be blamed in large measure on Galileo's polemical style, which seemed deliberately designed to cause outrage among his intellectual adversaries. He described their opinions variously as "childish" and "ignorant," while at the same time urging his own views with an immodesty that frequently was bellicose. He was, in a word, insufferable.

As always, Galileo had his eye on the main chance when he took his manuscript to his publisher, and he dedicated the book to his pupil of four years earlier and now the current occupant of the Medici throne in Florence, Cosimo II, fourth grand duke of Tuscany. But that was only the teaser. In a move calculated to gain him a plum patronage position in the Tuscan court he named the moons of Jupiter the Medician Stars. The dedication was as florid as anything in Baroque literature, complete with astrological references. After noting that the known constellations and planets had been named for Greek and Roman deities and mythological heroes he says it is only fitting that Cosimo should have the same honor.

> Indeed, the Maker of the stars himself has seemed by clear indications to direct that I assign to these new planets Your Highness's famous name in preference to all others. For just as these stars, like children worthy of their sire, never leave the side of Jupiter by any appreciable distance, so (as indeed who does not know?) clemency, kindness of heart, gentleness of manner, splendor of royal blood, nobility in public affairs, and excellency of authority and rule have all fixed their abode and habitation in Your Highness. And who, I ask once more, does not know that all these virtues emanate from the benign star of Jupiter, next after God as the source of all things good? Jupiter; Jupiter I say, at the instant of Your Highness's birth, having

> already merged from the turbid mists of the horizon and occupied the midst of the heavens, illuminating the eastern sky from his own royal house, looked out from that exalted throne upon your auspicious birth and poured forth all his splendor and majesty in order that your tender body and your mind (already adorned by God with the most noble ornaments) might imbibe with their first breath that universal influence and power. . . .

Galileo had smelled an opportunity to win the patronage of the Medici family, and he set out to reach that goal with characteristic impatience and bluntness. He wrote to a friend: "It is impossible to obtain wages from a republic [i.e., Venice], however splendid and generous it may be, without having duties attached. For to have anything from the public one must satisfy the public and not any one individual; so long as I am capable of lecturing and serving, no one in the republic can exempt me from duty while I receive pay. In brief, I can only hope to enjoy these benefits from an absolute ruler."[12]

Patronage would take him out of the university system with its wearisome academic politics and time-robbing teaching commitments. It would increase his income and his status. And most important of all, the identification with the powerful Medicis would provide a degree of insulation from the attacks of intellectual opponents, making it easier to publish controversial ideas. But patronage had its drawbacks, as well. Galileo would have value to the Medicis only insofar as he was seen to be a great discoverer of new things and a brilliant philosopher, the doyen of his profession. Patronage was a two-way street and the patron expected a payback. It typically came in the form of reflected glory—the more glory, the more dazzling the reflection. And so Galileo would have to perform. Furthermore, his "performance" would henceforth take place in the public realm rather than in the cloistered world of the university. Criticism of his work would become a matter for public consumption and debate, which would mean his rebuttals would have to be timely and convincing.

All these considerations would play a role in the increasingly tumultuous life he faced following his telescopic discoveries. Some commentators suggest that it was this pressure that pushed Galileo into his eventual confrontation with a Church that was actively seeking compromise. It would be, in any case, a far cry from the equable obscurity in which he had lived well into middle age.

In accepting the patronage of the Medicis and leaving the University of Padua, Galileo would be abandoning the republican regime of Venice (which ruled Padua) for the dictatorship of Tuscany. His fate would no longer be in the hands of a democracy, however rudimentary and inadequate, but would henceforth be subject to the whim of an absolute ruler. His old friend Giovanfrancesco Sagredo (whom he would later immortalize in his *Dialogue on the Two Chief World Systems*) expressed strong reservations: "Where will you find the same liberty as here in Venetian territory . . . where you had command over those who govern and command others? . . . If not ruined, you may be harried by the surging billows of court life and by the raging winds of envy. . . . Also your being in a place where the authority of the friends of Berlinzone, from what I am told, stands high, gives me cause for worry."[13]

Berlinzone, or Messer Rocco Berlinzone, was a wary nickname for the Jesuits. The Society of Jesus had been founded in 1540 and quickly became the dominant order within the Church of Rome and one of its chief instruments in implementing the strategies of the Counter-Reformation, Rome's response to the Protestant schism. Their leading role was due to their spectacular success. When the order was founded the Catholic Church was falling back in disarray on all fronts against the Protestant offensive. By 1615 Protestantism was staggering back, and it was the Jesuits more than any other factor who were responsible. Organized along military lines, Jesuits focused on education and missionary work, at which they excelled. They were fiercely loyal to Rome. The order's founder, Ignatius Loyola, told initiates, "We ought always to hold by the principle that the white I see I should believe to be black if the hierarchical Church were so to rule it."[14] Paradoxically, the order quickly gained a reputation for flexibility in adapting Catholic precepts to local custom not only in the far-flung corners of the world (like New France) where they established extensive missions, but in the courts of Europe. Some saw this as laudable; others called it devious and unscrupulous. In any case, Jesuits would be numbered among the leading humanist scholars in Europe by the end of the sixteenth century, and the young Galileo included himself among their admirers.

One of the main institutions of the Jesuit educational enterprise was the Collegio Romano, which Loyola himself had founded in 1551. It was here the order assembled its best minds, and in Galileo's time it may have been the best university in Christendom. The order's astronomers were second to none and thoroughly familiar with the theories of Copernicus, Tycho, and Kepler. When Pope Gregory XIII decided to reform the antiquated Julian calendar, it was to the Jesuits he turned, enlisting the top astronomer at the Collegio Romano, Christopher Clavius, with whom Galileo would later carry on a long correspondence. It is to Clavius that we owe the current Gregorian calendar. But it is worth noting that it was Clavius, as well, who wrote the introduction to the Council of Trent–authorized translation of the Bible known as the Clementine Vulgate, a cornerstone of Roman Catholic orthodoxy. And the same Clavius had served on the commission that tried and convicted Giordano Bruno in 1599 and had him burned at the stake for heresy. Adept though he might have been at the emerging new science, Clavius continued to see moral and natural philosophy as aspects of the same unified discipline, the one necessarily informing the other if knowledge was to be complete.

In general, and though they were loyally Aristotelian in their beliefs, the Jesuits rejected the idea of a necessary opposition between new learning and the established Church teachings, and thought rather of an accommodation or its fusion into a Christian humanism. Herein lay hidden the seeds of the eventual confrontation with Galileo, who had become a virulent critic of Aristotelian natural philosophy and who believed nothing short of its wholesale replacement was needed.

It was the Jesuit view that governments within the Christian commonwealth owed their very existence to the God-given authority of the pope, on whom all secular power ultimately depended. This view obviously opened the door to a possible denial of the authority of a secular prince, and that inevitably led to a distrust of Jesuits in the royal courts of Europe. The fact that the Society of Jesus was on the mind of Galileo's friend and adviser Sagredo may be explained by the fact that just three years before he wrote his letter to Galileo the Jesuits had been party to a tense clash of wills between the Venetian Senate and Pope Paul V. In 1605 the Senate had arrested two priests and charged them with serious crimes (one was accused of murder); Paul V demanded that they be turned over to Church authorities, and at the same time he ordered that Venetian laws restricting construction of new churches and monasteries be repealed. The Senate refused, citing the legal opinions of Father Paolo Sarpi, Venice's state theologian and a cogent critic of papal absolutism who is remembered as an early advocate of the separation of Church and state.

The pope, replying to Sarpi and the Senate with the carefully wrought legal arguments of Cardinal Roberto Bellarmine, the Vatican's senior theologian (head of state), went on to excommunicate the doge and other leading government officials and to place an interdict on all religious services in Venetian territory. The doge countered by ordering the local clergy to ignore the interdict, which they did, with the exception of the Jesuits and two older monastic orders. The Jesuits left Venice en masse, despite the Senate's warning that they would never be allowed back. Rumblings of military intervention on the pope's behalf by Spain and France were countered by Venice's threat to call on the Protestant states to its north for help, which led to fears that Venice would be lost to Protestantism. In the end, a face-saving truce was negotiated by a French envoy, Cardinal Joyeuse. Paul V's interdict had been a medieval anachronism no longer effective in the world of emerging nation-states. It was, in fact, the last time an entire state was to be placed under interdict by the Vatican.

Despite the truce mediated by Cardinal Joyeuse, ill feeling against Rome persisted in Venice, and this no doubt colored Sagredo's opinion when he wrote his "Berlinzone" warning to Galileo. But his concern that the Jesuits might interfere with the scientist's work was probably misplaced at the time. It would be many years before they turned against Galileo, and then it was not without provocation. In the interim, they were a powerful source of support, and he had a good opinion of them. It was the militant Dominicans, the "black-and-white hounds of the Lord," whom he initially had to fear.

Evidence for and against the Copernican system was seen to be of two kinds: physical, empirical evidence, on the one hand, and on the other, the wisdom of authority in the form of the writings of Church fathers like St. Thomas Aquinas and in Scripture. Galileo's astounding telescopic discoveries had all but eliminated the empirical, astronomical objections to the Copernican system that he now embraced with real, if guarded, enthusiasm. His observations of the orbit of Mars, the moons of Jupiter, the mountains of the Moon had all lent persuasive, though not entirely convincing, support to the heliocentric hypothesis. More such evidence was to come, and all of it would be confirmed by Jesuit astronomers. But lurking offstage, still unresolved and highly dangerous, were the theological objections. Though they at first glance seemed trivial in the face of the physical evidence, there was a deep, difficult aspect to them that lay waiting for the unwary like quicksand on a moor. As his adversaries in the Church lost one astronomical argument after another, it was to these objections that they would turn in their determination to resist what they saw as a threat to the foundations of Christian thought. Here, they held the high ground and they did not surrender it.

TEN

In the Medician Court • The Phases of Venus
Sunspot Disputes • Triumph in Rome
Cardinal Bellarmine

Two months after the publication of *The Starry Messenger*, Galileo visited his old friend and compatriot Belisario Vinta, the Florentine secretary of state, continuing his campaign for a court appointment. He followed up with a letter in which he spoke of the great plans he would be able to put in motion once relieved of his teaching obligations in Padua and of his desire to serve the grand duke:

> Particular secrets, as useful as they are curious and admirable, I have in great plenty. Their very abundance has worked to my disadvantage (and still does), for had I but a single one of these I should esteem it highly, and with that incentive I could have interested some great ruler, which I have not hitherto done or attempted. Great and remarkable things are mine, but I can only serve (or rather, be put to work by) princes, because it is they who are able to carry on war, build and defend fortresses, and for their regal sport make those great expenditures which neither I nor other private persons can. The works that I intend to bring to a conclusion are principally two books on the *System* or *Constitution of the World*, an immense conception, full of

> philosophy, astronomy, and geometry. Then there are three books *On Motion*, a science entirely new . . . and though other men have written on this subject, what has been done is not one-quarter of what I write, either in quantity or otherwise.[1]

Galileo tells Vinta he is prepared to accept the salary mentioned in their earlier meeting. But he asks for one further consideration: "But finally, as to the title of my position, I desire in addition to the title of 'mathematician' His Highness will annex that of 'philosopher'; for I may claim to have studied more years in philosophy than months in pure mathematics. As to my deserving the full title, their Highnesses may judge for themselves as soon as they give me an opportunity to deal in their presence with the men most esteemed in this subject."*[2]

In June 1606 Galileo resigned his position at the University of Padua, and in September of the same year he took up his new post as chief mathematician and philosopher to the Medicis. He soon had his telescope set up and pointed to the heavens, and more new marvels were revealed. On December 11 Galileo sent a letter to Kepler in Prague containing an anagram announcing an important discovery. This was an accepted method of protecting priority of discovery while official publication was pending. The anagram read: "*Haec immatura a me iam frustra leguntur o y.*" Kepler tried but was unable to unlock the puzzle and had to wait for Galileo to decode it in a letter a month later as "*Cynthia figuras aemulatur mater amorum*" or "The mother of love (Venus) imitates the appearances of Cynthia (the Moon)." Galileo was proclaiming his discovery that Venus goes through phases like the moon. This was news of the utmost importance. If, as Copernicus said, the Earth and Venus (along with the other planets) revolved around the Sun, the amount of the surface of Venus visible from Earth would have to vary according to the relative positions of the planets in their orbits. When Venus was on the opposite side of the Sun from the Earth, it should appear as a full disk. When it was on the same side of the Sun as the Earth, it should be dark, like a new moon. And

---

* Kepler, as Galileo was well aware, was simply chief mathematician to the Imperial court.

at intervening positions, it should display the various phases of fullness. To the naked eye, this did not appear to be the case. But through the telescope, Venus's phases were for the first time plainly visible.

The discovery, however, fell tantalizingly short of providing definitive proof of the Copernican system, since it could equally be explained within Tycho's planetary scheme in which all the planets orbited the Sun, but the Sun in turn revolved around the Earth. It was, however, a fatal blow for Ptolemy, because the pattern of phases observed by Galileo could not occur in the Ptolemaic geometry.

Galileo also discovered the rings of Saturn at this time, though he did not know it. He had seen what seemed to be two companion satellites very close by on either side of Saturn, which appeared and disappeared at intervals. What he was observing was the tilt of the planet's rings through a cycle in which they periodically are rim-on to the Earth and almost invisible. His own conjecturing on the strange disappearances was more melodramatic: "What can be said of so strange a metamorphosis? Were the two smaller stars consumed likes spots on the sun? Have they suddenly vanished and fled? Or has Saturn devoured his own children? . . . I cannot resolve what to say in a change so strange, so new, so unexpected."[3]

And indeed, who could have imagined rings around a planet?

Finally, as this excerpt from a letter to a friend indicates, Galileo had seen a series of dark blemishes moving across the face of the Sun. (He projected the Sun's image through a telescope onto a white screen, a technique suggested by one of his former students.) This was of course powerful new evidence against the Aristotelian idea of the incorruptibility of the celestial spheres. In this case Galileo was unable to claim priority of discovery. A Jesuit priest and scientist named Christopher Scheiner had seen the spots at the same time and documented his sightings in letters, which were eventually published under the pseudonym Apelles and brought to Galileo's attention by a mutual friend. (In fact, sunspots were certainly known in the time of Charlemagne and may have been mentioned by Virgil. Credit for first publication rightfully goes to Johann Fabricus of Wittenberg, in 1611, though this was apparently unknown to both Galileo and Scheiner.)[4] Scheiner suggested that the spots were swarms of hitherto unknown planets orbiting close to the Sun's surface.

Galileo, by now accustomed to having the field of discovery all to himself, reacted in a reply that was withering in its sarcasm and condescension, quite unnecessarily making a lifelong enemy of a competent scientific inquirer and potential ally. His wide-ranging pamphlet would be published (in 1613) under the title *Letters on Sunspots*, which is chiefly memorable for the fact that, in it, Galileo for the first time gives unequivocal public support to the Copernican system.* Scheiner continued his research and was eventually to come around to Galileo's position that the spots were on the surface of the Sun. It was he who would write the century's definitive book on the subject.

Galileo was aware that within weeks of his discoveries Jesuit astronomers at the Collegio Romano had begun observing the heavens with their own telescopes. One of the Jesuit astronomers, Giovan Paulo Lembo, had in fact independently noted the irregularity of the Moon's surface and the starry composition of the Milky Way and nebulae, using a telescope he'd built himself.[5] Throughout the summer of 1610 the Jesuits had tried to duplicate Galileo's observations of the moons of Jupiter, without success. Galileo himself had offered advice on the importance of properly mounting their instruments to avoid vibration and other accidental motion. In December 1610, using a new instrument of Lembo's, the Jesuits achieved confirmation.† Christopher Clavius wrote to Galileo immediately to inform him of this important corroboration.

The time seemed propitious for Galileo to journey to Rome and seize the initiative in what he had begun to see as his larger project: to popularize his discoveries among the educated elites and bring

---

*Printed under the auspices of the Lincean Academy, its full title is *Demonstrations Concerning Sunspots and their Phenomena, contained in three letters, written to the Illustrious Mark Wesler, Deemvir of Augsburg and Counselor to His Imperial Majesty*. As well as unequivocally endorsing the Copernican system for the first time in print, Galileo predicts it will soon be universally accepted.

†A six-month set of observations of the Jovian moons in Galileo's hand has survived among his papers. These are often attributed to Galileo, but it is now known that they were copied by Galileo from Jesuit observations, during his 1611 visit to Rome.

the powers within the Church to accept the significance he attached to them. This would eventually crystallize into a twofold ambition. The first phase was to crush the Aristotelian-Ptolemaic cosmology once and for all, replacing it with Copernicanism. The second was nothing less than to challenge the Church's thousand-year-old authority over the nature of truth. Galileo was increasingly convinced that he had discovered in his new science a superior approach to understanding the world. As a committed Catholic he was eager to find a way to allow it to flourish without needlessly damaging the Church. That meant finding a way for the coexistence of the empirical truth of Copernicanism, on the one hand, and the truth of revealed wisdom and Scripture on the other. As usual, he was confident that he had the answers, although on this visit he would remain uncharacteristically circumspect on the issue.

And so, in March 1611, a year after he had published *The Starry Messenger*, Galileo departed for Rome, arriving April 1 for a three-month stay in the Villa Medici with the family of the Tuscan ambassador. The city received him with a warmth and enthusiasm he must have found enormously gratifying after the chafing years in Padua.

Pope Paul V, for all his alleged anti-intellectualism and conservatism, granted Galileo a cordial audience in which he broke with protocol in refusing to allow the scientist to kneel. Cardinal Roberto Bellarmine, senior Vatican theologian, adviser to the Holy Office, examiner of prospective cardinals and a man destined for sainthood, also took an active interest in his discoveries. An ardent amateur cosmologist, Bellarmine was known within the Collegio for his public adherence to the "fluid heavens" theory, which had challenged the Aristotelian cosmology of crystalline spheres. Shortly after Galileo arrived in Rome the great theologian asked Christopher Clavius and the other mathematicians and astronomers of the Collegio Romano for their "sincere opinion" on the discoveries Galileo claimed:

> I know that Your Reverences are aware of the new celestial observations by a worthy mathematician using an instrument called a *canone* or *ochiale*. By means of this instrument even I have seen some very marvelous things concerning the moon and Venus, but I wish that you would do me the pleasure of telling

> me your sincere opinion of these things. . . . I want to know this because I hear various opinions spoken about these matters, and Your Reverences, versed as you are in the mathematical sciences, will easily be able to tell me if these new discoveries are well founded, or if they are rather appearances and not real.[6]

The Jesuit astronomers were in a position to reply promptly and unequivocally that the discoveries were real and had been confirmed by their own observations. How best to interpret them was a question they discreetly left untouched.

Galileo staged many demonstrations with his telescope while in Rome, some resoundingly successful, others less so.[7] On a clear spring night two weeks after being feted at the Collegio Romano, the astronomer entertained a group of eight interested observers on a raised clearing in the vineyards of a Monsignor Malvasia outside the Porta San Pancazio, the western portal through Rome's great wall. Clustered around the leather-bound telescope were a motley group, including Johann Scheck (also known as Terrentius), an advanced student at the Collegio Romano who would eventually travel to China as a missionary and there publish several books on astronomy and other subjects; Francesco Piffari of Siena, who had published an Italian translation of Clavius's astronomy text; Antonio Persio, a mathematician with Copernican leanings; Giulio Cesare Lafalla, a philosopher from Rome; and Johannes Desmiani, called el Greco, who was mathematician to a Cardinal Gonzaga. Of this gathering, only the Copernican Persio and the student Scheck would have been supporters of Galileo. The others were, in varying degrees, hostile. The chronicle reports that the men remained outside in the chill spring air listening to Galileo and by turns squinting through the telescope for seven hours, arguing among themselves about what it was that they had seen. In the end, as dawn broke, there was no agreement and an ill-humored parting.

Episodes such as this, though, were the exception to an overwhelmingly positive reception. The prestigious Lincean Academy of Prince Federico Cesi invited Galileo to join its elite membership, which had pledged itself "to fight Aristotelianism to the end." From the day of his acceptance he would proudly sign all his correspondence Galileo Galilei, Linceo. And perhaps most gratifying of all, the Collegio Romano's students of mathematics and astronomy organized a day-long symposium dedicated to celebrating his discoveries. A highlight was a commentary on *The Starry Messenger* presented by Odo van Maelcote, until the year previous the chairman of the Collegio department of mathematics. Maelcote declared that Galileo's discoveries and their confirmation by Jesuit astron-omers had demonstrated conclusively that the Sun was at the center of the planetary system. A surviving journal describes the event. "Galileo entered the grand hall of the Academy . . . and we, in his presence, expounded the new phenomenon before the whole university. . . . And we demonstrated with evidence, to the scandal of the philosophers, that Venus circles the sun. . . . With this public demonstration Galileo will return to Florence much reassured and, in a manner of speaking, with an honorary degree conferred by the universal consensus of this university."[8]

Before the year was out Clavius, mentor of his youth, now at death's door and still a dogged defender of the Ptolemaic system, would revise his famous textbook on astronomy to include full and generous mention of Galileo's discoveries. He advised readers to consult *The Starry Messenger*'s "careful and accurate" descriptions and concluded with the pregnant words: "Since things are thus, astronomers ought to consider how the celestial orbs may be arranged in order to save these appearances."[9]

ELEVEN

Bruno's Heresy • An Infinite Universe
Aristotle Revisited • The Struggles of Aquinas
The Return of Pythagoras

An ominous bronze statue of the philosopher and mathematician Giordano Bruno, broods over a little piazza in Rome called Campo dei Fiore.

Bruno was burned alive in the square on February 17, 1600, following seven years of imprisonment by the Inquisition. Galileo was thirty-six years old at the time, a newly installed professor of mathematics at Padua, father to a newborn girl named Virginia. Although Bruno was executed for doctrinal reasons rather than his scientific ideas, and although Galileo maintained long-standing friendships with at least two men who played important roles in his prosecution (Cardinal Roberto Bellarmine and Christopher Clavius), we can only assume that the incident was never far from Galileo's mind during his own encounters with religious authority.

The statue had been erected in 1887 thanks to a fit of pique on the part of an anti-clerical prime minister of a newly unified Italy, Francesco Crispi. The mayor of Rome of the time, Duke Leopold Torlonia, made the mistake of paying a visit to the cardinal vicar at the Vatican and asking him to convey the good wishes of the city to Pope Leo XIII on the occasion of the fiftieth anniversary of his ordination as a priest. Crispi instantly dismissed the mayor, and to reinforce his point raised the statue to Bruno like a middle-finger salute. Several other heretics are commemorated in medallions around the base: Erasmus, Vanini, Pallario, Sevetus, Wycliffe, Hus, Sarpie, and Campanella.

The brooding figure of Bruno presided over a lively food market where fruits and vegetables, meat and fish were being sold under brightly colored awnings and parasols. Many of the buildings enclosing the square had survived from his day. I could imagine the terrible scene as he was frog-marched to the stake, his tongue tied lest he cry out his heresies in public. I knew that the records of his trial by the Inquisition were kept in the same vault in the Vatican library as the Galileo documents, not accessible to the public, available only to select scholars. This was not an episode the Church was proud of, nor had it any reason to be.

Bruno was born in 1548 near Naples, the son of a professional soldier. After study at the University of Naples, he was ordained a Dominican priest, though even then his penchant for espousing unorthodox views caused some second thoughts within the order. Three years later a formal trial for heresy was prepared against him, for reading the banned works of Erasmus and publicly discussing the idea that Christ was not divine. He fled to Rome, where he was accused of a murder, and excommunication proceedings were begun against him. He fled again, this time abandoning the Dominican order. After wandering northern Italy he turned up in Geneva, where he converted to Calvinism. However, after publishing an attack on a prominent Calvinist professor he discovered that the Reformed Church was no more tolerant of his heterodox opinions than the Catholic: he was arrested and excommunicated. After abjuring his offending statements he was rehabilitated and allowed to leave the city. Following this he pursued the life of a wandering scholar, moving to Toulouse and then on to Paris, where he briefly adopted a sympathetic community of moderate Catholics before falling out with them. From there he made his way to London and Oxford and on to Germany, teaching and publishing wherever he went. In Helmstedt his publications earned him a second excommunication, this time by the Lutheran Church. By then he had made enemies from one end of the spectrum of Christian belief to the other.

In 1591 he accepted the invitation of a Venetian patrician,

Giovanni Mocenigo, to move to that city, which at the time was the most liberal of the Italian city-states. From there he went almost immediately to Padua, where he had heard there was an opening for chair of the department of mathematics. He applied for the post, but it was given instead to the young Galileo, whose views were considerably more orthodox. He returned to Venice and the hospitality of his aristocratic patron, only to be denounced by Mocenigo himself to the Venetian Inquisition for his heretical views. During the spring and summer of 1592 his trial was begun, but before it could be completed the Inquisition in Rome intervened with a demand for Bruno's extradition. In 1593 the seven-year ordeal of his imprisonment and trial by the Holy Office began.

At first, Bruno defended himself by arguing that he had no interest in theology, and that his views were principally philosophical (i.e., scientific). He was willing to concede some minor theological errors. He anticipated Galileo's argument that the Church should confine itself to issues of moral philosophy and theology and leave natural philosophy to philosophers and scientists. The Inquisitors did not accept this and demanded unconditional renunciation of his heretical theories. Bruno then mounted a second line of defense in which he argued that his views were not in conflict with the basic Christian tenets of God and Creation. This, too, was rejected. Finally Bruno protested that he would retract nothing and in any case no longer knew what retractions were expected of him. With that, Pope Clement VIII ordered his sentencing as an impenitent and obdurate heretic.

In the official newspaper report on his burning, the *Avvista di Roma* of February 19, 1600, emphasized what it called the capricious nature of Bruno's heretical teachings, which he had directed especially against the Holy Virgin and the saints. It alleges that he "obstinately wanted to die for them, and said that he was dying freely, as a martyr."

From the perspective of four hundred years after the event, it is possible to see (though not condone) the element of sheer exasperation on the part of the Church in this sad episode. Despite seven years of threats, browbeating, and reasoned argument, Bruno would not budge. He seemed fundamentally irrational to his Inquisitors, little short of a maniac. In the end, Bruno was burned not for any specific doctrine, though he subscribed to many that were heretical.* He was burned rather for his wanton *curiositas*, for his belief in the limitless capacity of man to know—to know, eventually, what God knows. The other side of that coin was a presumption that *all* knowledge was of a kind accessible to rational inquiry. And that in turn implied that what was not thus accessible was not knowledge but something else—superstition, perhaps.

Bruno was an early and vociferous exponent of the idea of continuous progress in intellectual endeavor. In his writings he pointed out that revered ancients such as Ptolemy had built their knowledge on the observations of their predecessors, who had in turn benefited from the achievements of others. Copernicus, he wrote, was in a better position than any of his predecessors to know the truth about the heavens because 1,849 years had elapsed since the early astronomer Eudoxus had employed still earlier Babylonian observations to develop his cosmology. As a matter of fact, said Bruno, in terms of intelligence, it is we who are the Ancients, and the age of Classical Greece belongs to the childhood of man. From this thought it followed that the Middle Ages—the name was coined in Bruno's time—were merely a provisional phase of human self-realization, a bridge on the way from antiquity to the comparative perfection of modern times.

Although Copernicus had not figured in the prosecution of Bruno, the great astronomer's ideas had set the stage for an issue that did—the question of the size of the universe. His postulating of an infinite universe had led Bruno directly to a rejection of belief in the Incarnation of God in Christ. It did not seem reasonable to Bruno that in an infinite universe God should have become incarnate only on Earth. In fact, the idea of the Incarnation made no sense at all, because the Divinity had already spent Himself, exhausted all possibilities, in creating the infinite universe. Infinitude implied that He had held nothing back. And so, nothing "supernatural" was possible.

If this strikes the modern reader as merely sensible, it is because

---

*The condemnation to burning was a deterrent, but it was also an act of faith in the ability of absolution granted before death to save the soul of the condemned. The Inquisition, having failed to convert the accused, invoked the supernatural power.

with Bruno, the Creator has become indistinguishable from a natural cause. Conventional Christian philosophy, which assumed a finite world, was able to state, as the theologian Bonaventura (1221–74) did, that while God had given away much of his treasure when He created the world, He had not given it all. Another great philosopher of the Church, William of Ockham (d. 1349), restated this as a belief that God is able to create something that He has not yet created and does not want to create. The Christian world was very different from the deterministic, proto-scientific universe of Bruno.

Bruno's philosophy of infinitude avoids the necessity of any notion of goals or ends (teleology) from creeping into discourse about ultimate causes. Instead of an intelligence at work in the universe, with Bruno we have a slot machine generating random choices based on nothing more than mathematical probabilities.

The further consequence of this is that in an infinite world, where possibilities for existence are infinite, there is no logic in the kind of special status both Aristotelianism and Christianity give to this particular planet and its neighboring astronomical bodies. In an infinite universe, there would logically be an unlimited number of solar systems, and, moreover, there was no reason to suppose that there would not be a similarly large number of inhabited planets. Even worse, from the point of view of the Church, was the unavoidable implication that there could be, and probably were, a great many, perhaps an infinite number, of universes. For there was nothing to prevent Bruno's infinitely powerful but impersonal God, constrained and impelled by logic, from reproducing universes ad infinitum. To the dyed-in-the-wool Aristotelians of the Church hierarchy, this seemed simply grotesque—an abomination, dangerous theological quackery.

Bruno's Inquisitors were not surprised when, unable to speak because of the leather gag, he defiantly twisted his head away from the bronze cross that was held before him as the faggots piled around him were lighted.

With this in mind, I continued reading Sister Celeste's monograph on Aristotle.

Even in the period of the collapse of Roman civil authority and barbarian invasions that we call the Dark Ages, the teachings of Aristotle had never been entirely lost to Western culture. His *Logic* was textbook material in the oldest of universities. But in the twelfth century a flood of works previously only rumored to exist began appearing, thanks in part to the communication with the Arab world fostered by the Crusades. These hitherto unknown works of Aristotle had been preserved not in the Latin libraries of Christian monasteries but by Arab scholars, and they reached Europe mainly as translations of Syrian and Arabic versions of the original Greek texts. They included all the philosopher's works on natural philosophy as well as his books on metaphysics, ethics, and psychology.

It would be difficult to overestimate the impact of this material. It was not a matter of a few diverse new texts appearing on the scene but rather the arrival of a complete, coherent, compelling worldview in competition with the time-worn traditional view.* It was an intellectual bombshell of almost unimaginable proportions.

Naturally, it was fiercely resisted. Traditionalists feared Aristotelian thought would shatter the coherence of the Western intellectual and religious tradition so painstakingly crafted and carefully nurtured over the previous thousand years. Ecclesiastical prohibitions abounded, and the teaching of Aristotle was forbidden at one university after another. As would be the case with the no less revolutionary thought of Copernicus and Galileo, however, the Church was remarkably lax and inconsistent in its enforcement of these strictures. As an institution, it seems to have recognized the inevitability of the success of this new philosophical paradigm, and sought to avoid a rout on the battlefield of ideas by staging a strategic withdrawal.

Coupled with the withdrawal would be a regrouping of intellectual resources under the auspices of the redoubtable Dominican friar Thomas Aquinas. As so many others were

---

*This was based mainly on St. Augustine's adaptations of Neoplatonist thought. At the time of the rediscovery of Aristotle its most radical expression was in the emergence of the new mendicant preaching orders, the Dominicans and the Franciscans.

doing, he had ignored Church prohibitions and found a teacher of Aristotle, in his case in Naples. His studies complete, his superiors moved him to the cockpit of medieval European philosophy and theology, the University of Paris.

Through a sustained and prodigious intellectual output over the years between 1268 and 1272 Aquinas succeeded in integrating Aristotle's thought into the corpus of Christian philosophy and theology. The price he paid was death from exhaustion. But by the late fourteenth century, far from being prohibited, the complete works of Aristotle had become required reading for students at the University of Paris, along with the more traditional fare provided by Plato and his followers from late antiquity, the Neoplatonists. Almost single-handedly, Aquinas had preserved the continuity of the mainstream European worldview for another four hundred years, until the ultimately successful challenge of Galileo and scientific realism.

Aquinas seemed to have understood that Aristotle's conquest was so swift and so convincing because he tapped into a vein of thought already emerging on its own in the West. It was a reaction against the asceticism and anti-materialism of the Dominicans and Franciscans. Their "contempt for the world" had seemed a natural consequence of the Platonic notion of an ideal world of pure forms or Ideas, accessible only to the intellect, which the objects of the natural world imitated with only imperfect success. We read in the excellent monograph of Josef Pieper, *A Guide to Thomas Aquinas*:

> Such denaturalization of the natural world sooner or later had to become intolerable; it is simply impossible to live a healthy and human life in a world populated exclusively by symbols. And by around 1200 the moment had come for Christendom, out of what may be termed a purely vital reaction, to grow sick and tired of seeing and denominating the world in that way. What the twelfth century lacked, and craved, was the concrete reality *beneath* this world of symbols. It was altogether logical that in the midst of the Christian West itself this irrepressible longing for the hard metal and the resistant substance of "real reality," so long

> submerged, must at last burst forth as a mighty, many-voiced and enthusiastic assent toward the Aristotelian cosmology, as soon as that whole complex of ideas about the universe hove in sight.[1]

I lifted the manuscript off my lap and placed it on my table at this point, surprised by what I had read. "What the twelfth century lacked, and craved, was the concrete reality *beneath* this world of symbols," Pieper had said. This seemed to me, to say the least, a questionable statement. It assumed that we had the option of living in a world of concrete, definitive reality. But Aristotle, for all his differences with his mentor and teacher Plato, still believed with him that ultimate reality was not accessible to human reason, and that what we conceive of as the real world is merely a reflection, a shadow of something beyond. The "reality" beyond the world of material things was exactly what medieval Christianity had sought to discover both before and after its encounter with Aristotle. It seemed to me that Pieper had projected a twentieth-century materialism back on the world of the twelfth century. I glanced ahead in the manuscript—Sister Celeste made no comment on the quotation. Perhaps it seemed obvious to her that the thirteenth century's "vital reaction" in favor of materialism, if it had happened at all, was fundamentally misguided and therefore no comment was needed. I carried on reading.

> What the sudden reappearance of Aristotle did was bring to a boil the long simmering debate over the apparent dichotomy between two approaches to knowledge—faith and reason.*

---

*The traditional relationship between faith and reason was stated succinctly and elegantly by Augustine in his "Letter to Consentius":

> Heaven forbid that God should hate in us that by which he made us superior to the other animals! Heaven forbid that we should believe in such a way as not to accept or seek reasons, since we could not even believe if we did not possess rational souls. Therefore in certain matters pertaining to the doctrine of salvation that we cannot yet grasp by reason—though one day we shall be able to do so—faith must precede reason and purify the heart and make it fit to receive and endure the great light of reason; and this is surely something reasonable. Thus it is reason-

Since both had their source in God, it seemed they could not be contradictory. St. Anselm, the founder of medieval Scholastic philosophy, thanked God that "what by thy gift I first believed, I now by thy illumination understand, so that even though I refused to believe in thy existence, I could not fail to grasp it by intelligence."[2]

But this raised a troublesome question, one that would emerge four hundred years later with devastating force through Galileo and the scientific thinkers who followed him: if the two approaches to knowing were equivalent to each other in terms of the security or reliability of the knowledge they provided, what was the point of religion?

Aquinas needed to find a solution that would accept the legitimacy of reason while at the same time protecting the sovereignty of religious experience. Already, many of his contemporaries at the University of Paris were ready to place Aristotle's philosophy on a par with, or even above, the teachings of the Church fathers,* the scholarly theologians who had carefully constructed the body of Christian thought over the previous millennium. Aquinas, though a great admirer of Aristotle,

---

able for the Prophet [Isaiah 7:9] to have said: "unless you believe, you will not understand." Here he was doubtless distinguishing between these two things and advising us first to believe, so that afterwards we might understand what we believe. It is thus reasonable to require that faith precede reason. . . . If, therefore, it is reasonable for faith to precede reason in certain matters of great moment that cannot yet be grasped, surely the very small portion of reason that persuades us of this must precede faith. (Quoted in David C. Lindberg, "Science and the Early Church," in *God and Nature: Historical Essays on the Encounter between Christianity and Science*, eds. David C. Lindberg and Ronald L. Numbers [University of California Press, 1986], 27–28.)

*A community of Aristotelian scholars had sprung up in Paris that advocated a system of thought later commentators have called "philosophism." Like the scientism that it foreshadowed, it was jealously exclusive in its understanding of truth, dismissing as unnecessary and irrelevant both the knowledge to be gained through Christian revelation (or other spiritual insight) and the scriptural writings based on that revelation. Philosophy, they decided, was autonomous of religion and capable of establishing ultimate truth in its own light, and those things not capable of being investigated by reason were not to be believed. This of course left little or no place for faith in human affairs.

maintained a critical stance that pleased neither the Church conservatives nor the radical so-called philosophists of the university. He drew a firm line between reason and revelation, between nature and grace. "The argument from authority is of all arguments the weakest," he declared. "What is well said by the ancients we will accept for our profit; what they said wrongly we will discard . . . [T]he aim of philosophy is not to know what men have thought, but how the truth of the matter stands."

Aquinas accepted the proposition that true knowledge obtained through reason on the one hand and revelation on the other could never contradict each other, but he drew some important distinctions. Reason, because it begins with self-evident principles and proceeds by rigorous rules of logic toward its conclusion, provides what might be termed a more precise knowledge than religion. It is, to put it another way, a more instrumentally useful knowledge; it can be used to get things done. The knowledge provided by religion, on the other hand, because of its source in divine revelation, is more secure, less vulnerable to contradiction or erosion over time. It is knowledge that can be counted on. Therefore faith speaks with greater authority and certainty than reason. Furthermore, Aquinas pointed out, some areas of knowledge are completely beyond the grasp of reason and the scientist. For example, the nature of the Trinity and the Incarnation are accessible only through faith. This was not because such knowledge is by its nature denied to human understanding, but because in our present condition of imperfection we cannot grasp it. For the redeemed in heaven, such questions are every bit as self-evident as the simplest of logical syllogisms.

On these distinctions between the two types of knowledge, Aquinas rests his definition of the difference between theology and religion on the one hand and science and metaphysics on the other. The theologian begins with the revealed wisdom of God as found, for instance, in Scripture and works downward to nature, the world of God's creation. Scientists (and the philosopher) start from nature, which is the proper object of human reason, and work upward through logical demonstration to

God. Religion, in other words, begins with the certitude of God's existence and occupies itself with the problem of why things are as they are, while science interests itself in the general character of things.

On the ancient question of *curiositas*, or the realm of legitimate curiosity, Aquinas accepted Aristotle's dictum that man is by nature a curious animal and that curiosity is on the whole an admirable trait. There can never be too much knowledge of the truth, Aquinas said, however, too great an emphasis on the drive to know can be harmful. This was an acknowledgment of Augustine's deep reservations about *curiositas*. Aquinas had this to say about maintaining a balance: "As regards knowledge there is a tension of opposites; the soul has an urge to know about things, which needs to be laudably tempered, lest we stretch out to know beyond due measure; while the body has an inclination to shirk the effort of discovering. As for the first, studiousness lies in restraint and economy, and as such is counted a part of temperance. As for the second, it is praised for a certain eagerness in getting to a conclusion. . . ."[3]

Aquinas gave to astronomy a special status in his consideration of *curiositas*, in recognition of the Aristotelian view that the heavens were of a fundamentally different nature from the sub-lunar world. In his commentary on Aristotle's *On the Heavens*, Aquinas dismisses inquiry into certain cosmological questions as fruitless. On the question of whether the sphere of the stars moves, or is in fact stationary while the Earth rotates, giving the stars the appearance of motion, he states: "If someone exerts himself to make assertions about such difficult and obscure matters and wants to assign them causes, and thus claims to extend his inquiry to everything and to omit nothing, then this must be regarded either as an indication of great stupidity, from which arises his inability to distinguish between the accessible and the inaccessible, or as an indication that he proceeds with great thoughtlessness and ultimately presumption, from which it appears that the man does not correctly evaluate his ability in regard to investigation of the truth."[4]

Here, in Aquinas, was what could have been an explicit reference to Galileo were it not for the fact that Galileo would not be born for another three hundred years. Here was the script that would be followed by the Church reactionaries whose triumph in the events of 1633 would have such devastating impact on the Church they thought they were protecting.

I continued to read.

> But Aquinas appears ambivalent here. He immediately amends his position, saying that whether or not a man should be condemned for pursuit of fruitless speculation in this way depends on his motive. If his motive is love of truth, he may be forgiven. But if he is interested rather in showing off his own abilities, he should not be. Furthermore, he says, there are times when extraordinary statements about areas of apparently inaccessible knowledge are not simply vain speculations but genuine insights, the fruits of a superior intellect: "If someone has gone further and achieved greater confidence in knowledge of the essential causes of these phenomena than seems possible according to the usual measure of human knowledge, then thanks rather than blame is due to such discoverers of necessary causal connections."[5]

Here, more than almost anywhere else in his writings, you could feel Aquinas struggling to integrate two fundamentally incompatible approaches to knowledge into the same philosophy. Here, in a few words, a single sentence, was the essence of Galileo's defense in his struggle with Church authority. And yet it was this same body of remarkably equivocal theology that the Church would rely on in his condemnation.

I knew that Aquinas had compared the lust for what we would call speculative scientific knowledge with envy of another's possession; an envy of God's exclusive knowledge.[6] It is, he said, an expression of fallen man's sorrow over not being God Himself. It was Bruno's sin. The self-indulgent pleasures of this quest for knowledge for its own sake, for piling fact upon fact, could be explained as willful, extravagant compensation for what humans have found they are unable to reach. It is a kind of pact with the devil. It is—of course!—the curse of Doctor Faustus.

Was it Galileo's sin as well, I wondered? In Christopher Marlowe's *Doctor Faustus*, the magician sells his soul to the devil in return for twenty-four years of additional life during which he will be granted every pleasure and access to all knowledge. But in Galileo's case an opportunity to repent was offered. Galileo accepted it and was saved. Or did he? In the myth his repentance is insincere. "Still, it moves," he whispers, after he has been compelled to read his official abjuration of Copernicanism. But as the historian Stillman Drake, a leading authority on Galileo, has shown, this episode is an invention of the eighteenth-century Italian author Giuseppe Baretti. The story's first appearance in print is in Baretti's *Italian Library*, published in London in 1757. In Baretti's version, as soon as he was released by the Inquisition, Galileo looked up to the sky and down to the ground, stamped his foot, and uttered the famous words. "[The story] was quickly picked up by other writers, who generally made it appear that Galileo had said these words as he rose from his knees after abjuring before the Inquisition. This preposterous version caused most serious writers to reject the whole story as a myth created to fit Galileo's personality rather than the truth." Drake goes on to conclude, rather unnecessarily, that it is possible Galileo made the statement at some time after he left Rome. It would be remarkable if he had not.[7]

Galileo was no Doctor Faustus, I concluded. I returned to the nun's paper.

> In these ways Aquinas hoped he had secured the leading position of theology in the realm of knowledge while at the same time accommodating the scientific approach to inquiry represented in Aristotle.
>
> Aristotle had shown Aquinas how to redraw the lines connecting nature and divinity in high relief, and make legitimate the study and appreciation of nature and its wonders. The Platonic view of ideal forms was not entirely jettisoned, but Aquinas, following Aristotle, incorporated the pure form or Idea *into the thing itself*—the thing and its ideal form were no longer separated by an unbridgeable gap as they had been in Plato and in earlier Christian thought. Pythagoras thus had made a sudden and unannounced return to mainstream thought, and this

opened the way for Galileo and materialist thinkers who followed him to make this crucially important argument: the "ideal form" that was the defining essence contained within the thing itself was properly represented by the purity of mathematics, and therefore mathematics in some sense *defined* the things in nature. Or as Galileo would put it, nature is written in the language of mathematics.

It has been argued with validity that the fatal flaw in Aquinas' theology was that he drew too stark a distinction between knowledge gained on the one hand by faith and on the other by reason. As the science he had helped to inaugurate matured over the centuries, it would become clear that knowledge obtained by science, for all its debt to reason, is no less indebted to faith. For, ultimately, scientists base their knowledge on faith at several levels: faith in the reliability of mathematics; faith in the essential order of the universe; faith that similar causes will be followed by similar effects; faith in the validity of extrapolation from conceptual models to the "real" world of nature; and finally faith in the very existence of such a real world. In each of these areas, a comfortable certitude cannot be obtained through reason, and is accessible only by faith. To the extent that a foundation in faith defines religion, science is every bit as much a religion as Christianity.

In its essence, the faith of modern science is little different than that of the ancient Pythagoreans. It boils down to a belief that number is an inherent property of the form of things, to a faith that number is in an important sense what things *are*. The opposing position, the one taken by many modern philosophers of science, is that number is not inherent in things as Aristotle and Aquinas said, nor is it the defining nature of the pure forms of Plato. Rather, number is imposed on things by us, as a way of understanding them. Number does not occur naturally in the world, but is a creation of the human mind, a template we impose on nature in order to better comprehend her. A religious view of this might be that number is our imperfect way of understanding the mind of God, as a map is used to understand the geography of an imperfectly explored continent. But the map, it must be remembered, is not the territory.*

The nun's manuscript ended with this suggestive sentence, scrawled in red ink: "In the modern world, then, what chiefly distinguishes science from religion is the matter of *success*."

---

* St. Augustine said, "Faith is to believe what we do not see, and the reward is to see what we believe."

TWELVE

Storm Warnings • Cardinal Barberini
The Conflict Is Joined: "Letter to Christina"
Bellarmine's Response

As the authorities in Rome began to fully understand the implications of the astronomical discoveries Galileo had described in his *Starry Messenger*, and more especially his interpretation of them, debate turned increasingly to their implications for the traditional teachings of the Church. This was a field in which discretion was always the better part of valor and never more so than in Galileo's time, with the Counter-Reformation in full swing, the Jesuits at the height of their power and influence, and the Dominicans doggedly vigilant for any whiff of heresy. Add to that the agony of the wars of religion that were devastating northern Europe (1618–48) and the recent doctrinal pronouncements of the Council of Trent (1545–63)*

*In response to the Protestant Reformation, this council, held at Trent in northern Italy, clarified and redefined the Church's doctrines. The council's reforms and doctrinal canons were the basis of the Counter-Reformation and became definitive statements of Catholic belief. On the key question of scriptural interpretation it decreed that "no one relying on his own judgment and distorting the Sacred Scriptures according to his own conceptions shall dare to interpret them contrary to that sense which Holy Mother Church, to whom it belongs to judge of their true sense and meaning, has held or does hold, or even [to interpret them] contrary to the unanimous agreement of the Fathers. . . ."

and the result was an atmosphere that was anything but conducive to the kind of radical theological innovation Galileo was promoting.

Nowhere outside Rome was discussion of Galileo's discoveries and opinions more frequent and animated than in the Poggio Imperiale, not far from Galileo's villa on the southern outskirts of Florence. The Poggio was the home preferred by Grand Duke Cosimo II and his young wife, Maria Madeleine of Austria, and it was here that Galileo must have announced his latest discoveries to his patrons. Often, the scientist would be invited for dinner, when he would be expected to participate in disputations with other learned men for the entertainment and edification of his hosts and their other guests. On one such summer evening about the time *The Starry Messenger* was published, talk turned to Galileo's latest theories on specific gravity and how solid bodies stay afloat in water.

At the table were two cardinals of a scientific bent. One of the clergymen, Cardinal Gonzaga (whose mathematician would later partake of the long and inconclusive night of observations outside the Roman wall), upheld the Aristotelian position in opposition to Galileo. The other, Cardinal Maffeo Barberini, a native of Tuscany, energetically joined the debate on Galileo's side.

The occasion is noteworthy for two reasons. First, the grand duke, captivated by Galileo's ideas, urged him to publish them forthwith. As a result Galileo spent much of late 1611 and early 1612 involved in this project, which resulted in the little book *Discourse on Floating Bodies*. It was a command performance, an excursion into physics for which Galileo apologized to readers who were awaiting his promised opus on astronomy and world systems.

In it, Galileo's rhetorical style reached new heights. He systematically set out his opponents' positions, augmenting them with new arguments of his own, only to demolish the whole structure with the results of his experiments. As the historian Stillman Drake observes, "It was a device he was to employ extensively in his later works, and one which accounts for his vast influence with nonprofessional readers as well as his extreme unpopularity with the targets of his polemical compositions."[1] Also important is the fact that Galileo chose, as he had with *The Starry Messenger*, to write in Italian rather than in Latin, which was the traditional language of the mathematician and philosopher. In doing so he was deliberately appealing to a broad public for support for his anti-Aristotelian project, once again to the disgust of his more conservative opponents, for whom this was an unfair and even dangerous tactic. But Galileo was insistent in his populist approach. His rationale clearly foreshadows the outlook of the approaching modern age:

> I notice that young men go to universities in order to become doctors or philosophers or anything so long as it is a title and that many go in for those professions who are utterly unfit for them, while others who would be very competent are prevented by business or their daily cares which keep them away from letters. Now these people, while provided with a good intelligence, yet, because they cannot understand what is written in *baos* [academic Latin], retain through life the idea that those big folios contain matters beyond their capacity which will forever remain closed to them; whereas I want them to realize that nature, as she has given them eyes to see her works, has given them a brain apt to grasp and understand them.[2]

The second reason that this particular dinner-table debate has been preserved in history is that Barberini, the cardinal who sided with Galileo, would become Pope Urban VIII and in that position would play a decisive role in Galileo's disastrous clash with the Church. Barberini left the dinner an enthusiastic admirer of Galileo, to whom he wrote a eulogizing poem. The two were to remain on friendly terms right up to the climactic events of 1633.

It wasn't only the evening meal that occasioned scientific and philosophical discussion at the Poggio—breakfast could sometimes be just as stimulating. Such was the case one morning in December 1613. This time Galileo was absent, recovering at the villa of a friend from one of his frequent bouts of undefined illness. But we have a detailed description of the event in the hand of one of his students, Benedetto Castelli, who had just taken over the chair of mathematics at the University of Pisa. The grand duke, educated in science and mathematics by Galileo, took a personal interest in the academic staff of universities in Tuscany, and young Castelli, a Benedictine monk, had been invited to the ducal palace for the ritual once-over. Just as interested, though for different reasons, was the grand duke's mother, the dowager Duchess Christina of Lorraine, a woman of great piety and a matriarchal figure of influence in the courts of three generations of Medici rulers. On December 14, 1613, Castelli wrote to Galileo:

> Wednesday morning I was dining at the court when I was asked about the university by the Grand Duke. I gave him a detailed account of things, with which he showed himself well satisfied. Then he asked me if I had a telescope, to which I replied yes, and fell to talking of my observations of the Medician planets made the previous night. Madame Christina wanted to know their position, and thereupon the talk turned to the necessity of their being real objects and not illusions of the telescope. Their Highnesses asked Professor Boscaglia* about this and he replied that their existence could not be denied. I then contributed all that I knew and could tell them about Your Excellency's wonderful discovery and establishment of the orbits of these planets. Don Antonio de' Medici, who was present at the table, beamed at me and showed himself well pleased by what I had said. After much talk, which went off quite solemnly, [the meal] was finally over and I left. But I had hardly come out of the palace when Madame Christina's porter overtook me and told me that she wished me to return. Now before I tell you what ensued, you must first know that while we were at table Dr. Boscaglia had the ear of Madame for a while; and, conceding as true all the new things you have discovered in the sky, he said that only the motion of the earth had something incredible in it and could not take place, in particular because the Holy Scripture was obviously contrary to this view.
>
> Now, getting back to my story, I entered into the chambers of Her Highness, and there I found the Grand Duke, Madame Christina and the Archduchess [Maria Madeleine], Don Antonio, Don Paolo Giordano [Orsini], and Dr. Boscaglia. Madame began, after some questions about myself, to argue the Holy Scripture against me. Thereupon, after having made suitable disclaimers, I commenced to play the theologian with such

---

*Cosimo Boscaglia, professor of philosophy at the University of Pisa.

> assurance and dignity that it would have done you good to hear me. Don Antonio assisted me, giving me such heart that instead of being dismayed by the majesty of Their Highnesses, I carried things off like a paladin. I quite won over the Grand Duke and his Archduchess, while Don Paolo came to my assistance with a very apt quotation from the Scripture. Only Madame Christina remained against me, but from her manner I judged that she did this only to hear my replies. Professor Boscaglia never said a word.[3]

The letter must have both amused and unsettled Galileo. Aware as he was of the dangers inherent in meddling in theology in these contentious times, it could not have given him much comfort to have been represented in discussions of the most sensitive kind by a neophyte, however talented. He immediately set about composing a long and impassioned letter that was to become his testament on the proper relationship between science and religion, a document of great historical significance.

Initially sent to Castelli, it was widely circulated in copies, one of which fell into the hands of a Dominican friar named Tommaso Caccini, who, in December 1614, preached a sermon from his pulpit in Florence against mathematicians in general and Galileo in particular. In February 1615 another Dominican named Lorini filed a written complaint against Galileo with the Holy Office in Rome, and a month later Caccini appeared in person before the Inquisitors to charge Galileo with suspicion of heresy. He based his accusations on the *Sunspot Letters*, which Galileo had published in 1613, the letter to Castelli, and on hearsay testimony of two informants. The Inquisition investigated the charges and conducted interviews with Caccini's supposed "witnesses" but found the allegations baseless, and no further action was taken.

The investigators did, however, find it necessary to request a formal opinion on Copernicanism from a group of eleven theological consultants. With that, the inexorable judicial processes that would have such enormous repercussions had been set in motion.

All this was done in secret, as was the Inquisition's practice, but rumors that something was afoot reached Galileo's ear. He decided on a bold course of action: he would withdraw the letter to Castelli from circulation and replace it with another, which revised, strengthened, and amplified his arguments. Then he would present his case in person, in Rome. The revised letter he addressed to his patron's mother, a more suitable recipient than Castelli for such an important statement of position, and a safe one. It became the famous "Letter to the Grand Duchess Christina."

In his essay, Galileo responds to charges that Copernicanism must be wrong, even heretical, because it seems to contradict specific passages of Scripture. These passages refer, always obliquely, to the Earth's stationary position in the universe, and to the movements of the Sun.* Galileo begins with a straightforward assertion of his belief in the truth of the Copernican hypothesis. His critics are aware, he says, that

> on the question of the constitution of the world's parts, I hold that the Sun is located at the center of the revolutions of the heavenly orbs and does not change place, and that the Earth rotates on itself and moves around it. Moreover, they hear how I confirm this view not only by refuting Ptolemy's and Aristotle's arguments, but also by producing many for the other side, especially some pertaining to physical effects whose causes perhaps cannot be determined in any other way, and other astronomical ones dependent on many features of the new celestial discoveries; these discoveries clearly confute the Ptolemaic system, and they agree admirably with this other position and confirm it.

Galileo implies repeatedly that the truth of the Copernican hypothesis has been thoroughly and completely demonstrated. He then goes on to argue, with strong support drawn from the writings of St. Augustine, that Scripture is frequently allegorical (as when it

---

*One such passage occurs in Joshua, where God commands the Sun and Moon to stand still over the valley of Ajalon to allow the Israelites to vanquish their foes (Joshua 10:12–13). In Ecclesiastes it is stated: "the earth abideth forever," and "the sun also rises, and the sun goeth down, and hastens to his place where he arose." Psalms 19:5 contains the passage: "[the Sun] exults as a strong man to run his course. . . ."

attributes human physical features to God) and often needs interpretation in order to draw out its true meaning.* That being the case, he says, it does not make sense to insist that the Earth is stationary and the Sun moves around it simply because this is implied in a handful of scriptural references. If the physical evidence points to the Earth's movement around the Sun, then Scripture must be reinterpreted to coincide with that evidence. Interpreting the Bible is a very serious undertaking, he says, and to do so without consideration of all the physical evidence available is to act irresponsibly. Here he quotes Augustine at some length, and to good effect:

> Usually [states Augustine], even a non-Christian knows something about the earth, the heavens, and the other elements of this world, about the motion and orbit of the stars and even their sizes and relative positions, about the predictable eclipses of the sun and moon, the cycles of the years and the seasons, about the kinds of animals, shrubs, stones, and so forth, and this knowledge he holds to as being certain from reason and experience. Now it is a disgraceful and dangerous thing for an infidel to hear a Christian, presumably giving the meaning of Holy Scripture, talking nonsense on these topics, and we should take all means to prevent such an embarrassing situation, in which people show up vast ignorance in a Christian and laugh it to scorn. The shame is not so much that an ignorant individual is derided, but that people outside the household of faith think our sacred writers held such opinions, and, to the great loss of those for whose salvation we toil, the writers of our Scripture are criticized and rejected as unlearned men. If they find a Christian mistaken in a field which they themselves know well and hear him maintaining foolish opinions about our books, how are they going to believe those books in matters concerning the resurrection of the dead, the hope of eternal

* References from Augustine are believed by modern authorities to have been provided to Galileo by Castelli and other friends with strong theological backgrounds. Galileo seems to have had a fairly superficial understanding of them, and often extends Augustine's arguments well beyond where the saint would have agreed to take them.

> life, and the kingdom of heaven, when they think their pages are full of falsehoods on facts which they themselves have learnt from experience and the light of reason?[4]

In any case, Galileo argues, the Bible is not a scientific text. It is a moral and spiritual document and should not be relied on for information on the nature of the physical universe. Certainly, the fact that it may appear to contradict evidence produced by scientific inquiry should not be enough to debunk the scientific claims. Moreover, Galileo insists, Scripture should not be used to deny any scientific claim that is *susceptible* to confirmation, even if the confirmation may not yet have been achieved. How it is to be determined that a claim *may* be proved, in the absence of that proof, is a problem he does not address. (However, Galileo clearly believed that there was little that science could not illuminate. He would explicitly state later in his *Dialogue on the Two Chief World Systems* what he only implies in the "Letter to Christina"—that while man's knowledge of the world could never be as *extensive* as God's, certainty nevertheless could be attained in a continuously expanding range of specific *intensive* areas of study. "I believe," he was to say, speaking through his alter ego, Salviati, "that [intensive] knowledge equals the Divine in objective certainty, for it is able to comprehend necessity, above which it is not possible to have greater certainty.")[5]

This left the role remaining to religion as one of filling in the gaps—temporarily providing explanations in areas into which science had not yet expanded its knowledge. Galileo does not appear to have seen this implication, but the Church clearly did, and found unacceptable the inevitable marginalization it implied.

The hard kernel of Galileo's argument against allowing the authority of Scripture to influence science is this: the Bible is the Word of God and as such cannot err. However, interpreters of the Bible are human, and thus make mistakes of interpretation. Even the pope may make such mistakes, in matters that are not directly related to faith, such as the movements of the planets.

"Certainly," Galileo says, "no one doubts that the Supreme Pontiff has always an absolute power to approve or condemn, *but it is not in the power of any created being to make things true or false, for this belongs to their own nature and to the fact* [emphasis added]."[6]

Nature is also the revelation of God, he continues, but *nature is its own interpreter*, and never errs. In his words: "But nature, on the other hand, is inexorable and immutable; she never transgresses the laws imposed upon her, or cares a whit whether her abstruse reason and methods of operation are understandable to men. For this reason it appears that nothing physical which sense-experience sets before our eyes, or which necessary demonstrations prove to us, ought to be called in question, much less condemned, upon the testimony of Biblical passages. . . ."[7] He concludes that certainties arrived at in physics, therefore, ought to be used to help interpret Scripture. But the reverse is not true.

Here, in concise form, is what I have characterized as "Galileo's mistake." It is an error that has been understood by philosophers from the eighteenth century onward, from David Hume to Immanuel Kant to Thomas Kuhn, with increasing clarity. The mistake is in the belief that nature is its own interpreter. It is not. Nor is it the case, as Galileo claimed, that "it is not in the power of any created being to make things true or false." It is simply not correct to assert, as Galileo did, that there is a single and unique explanation to natural phenomena, which may be understood through observation and reason, and which makes all other explanations wrong.

Scientists do not discover laws of nature, they invent them. Scientists do not observe "nature in the raw"—their observations are filtered through layers of subjective impression and social conditioning. Scientific "facts" about nature are not preexisting truths, they are human constructs, the products of human minds. The models that scientists build to represent what they see in nature are not literal representations of nature, but analogies, metaphors, simulacra. The truth that science "discovers" is not objective and immutable, it is subjective and socially contingent. That is why, from time to time, there are "revolutions" in science that overthrow one complete set of assumptions in favor of another, in the way that Galileo and Newton overthrew Aristotle, and Einstein, in turn, overthrew Newton.

These revolutions are relatively few and far between because science is a highly conservative enterprise, as it ought to be. It wants to make sure that when it changes its mind, it does so with good reason. Religion is conservative for precisely the same reason, as Galileo would soon discover.

Characteristically, and in spite of the sensitivity of the subject matter, Galileo minces no words in making his case to Christina. He is contemptuous and dismissive of those with whom he disagrees. "Officials and experts in theology should not arrogate to themselves the authority to issue decrees in professions they neither exercise nor study," he asserts. Aristotelian philosophers are attacked as "superficial and vulgar writers" of "simulated" and "insincere ... religious zeal." Though the letter is addressed to a dowager duchess of a certain age, and is a response to questions she has raised with Castelli, he quotes a passage from St. Jerome that can hardly have been expected to please her: "The garrulous old woman, the doting old man, and the wordy sophist, one and all take in hand the Scriptures, read them in pieces, and teach them before they have learned them. Some with brows knit and bombastic words, balanced one against the other, philosophize concerning the sacred writings among weak women. Others—I blush to say it—learn from women what they are to teach to men. . . ."[8]

The "Letter to Christina" was not to be published for many years, for reasons that will become evident.* As Galileo completed writing it, the committee of consultants in Rome were drafting their assessment of Copernicus, and rumors in Rome and Florence were now rife that the Holy Office was about to act, that Copernicanism was to be banned and that Galileo himself would suffer some form of censure. Nevertheless, his friend Piero Dini wrote to him from Rome with reassuring news about a meeting he had had with the Vatican's chief theologian:

> With [Cardinal Roberto] Bellarmine I spoke at length of the things you had written. . . . He said that as to Copernicus, there is no question of his book being prohibited; the worst that might happen, according to him, would be the addition of some material in the margins of the book to the effect that Coperni-

---

*The "Letter" was to be included with a Latin translation of the *Dialogue on the Two Chief World Systems* that was published in Germany in 1635. The translation arrived late at the printers and had to be printed separately, in 1636.

> cus had introduced his theory in order to save the appearances. . . . And with a similar precaution you may at any time deal with these matters, If things are fixed according to the Copernican system [he said] it does not appear currently that they would have any greater obstacle in the Bible than the passage "[the Sun] exults as a strong man to run his course [Psalms 19:5]" etc., which all expositors up to now have understood by attributing motion to the sun. And although I replied that this could be explained as a concession to our ordinary forms of expression, I was told in answer that this was not a thing to be done in haste, just as the condemnation of opinions was not to be passionately hurried.[9]

Bellarmine also provided a more direct reaction to Galileo's "Letter to Christina" in comments that are now taken to be the most authoritative statement of the Church's opinion at the time they were written, in April 1615:

> [T]o say that assuming the earth moves and the sun stands still saves all the appearances better than eccentrics and epicycles is to speak well. This has no danger in it, and suffices for mathematicians. But to wish to affirm that the sun is really fixed in the center of the heavens and merely turns upon itself without traveling from east to west, and that the earth is situated in the third sphere and revolves very swiftly around the sun, is a very dangerous thing, not only by irritating all the theologians and scholastic philosophers, but also by injuring our holy faith and making the sacred Scripture false. . . . Secondly, as you know, the Council [of Trent] would prohibit expounding the Bible contrary to the common agreement of the Holy Fathers. . . . Third, I say that if there were a true demonstration that the sun was in the center of the universe and the earth in the third sphere, and that the sun did not go around the earth but the earth went around the sun, *then it would be necessary to use careful consideration in explaining the Scriptures that seemed contrary, and we should rather have to say that we do not understand them than to say that something is false which had been proven. But I do not think that there is any such demonstration, since none has been*

> *shown to me.* To demonstrate that the appearances are saved by assuming the sun at the center and the earth in the heavens is not the same thing as to demonstrate that in fact the sun is in the center and the earth is in the heavens. I believe that the first demonstration may exist, but I have very grave doubts about the second; and in case of doubt one should not abandon the Holy Scriptures as expounded by the Holy Fathers [emphasis added].[10]

The message to Galileo from the highest theological authority could not have been plainer: unless and until you are in a position to clearly and definitively demonstrate the motion of the Earth and the stability of the Sun, keep your own counsel.

THIRTEEN

Under Attack • A Dangerous Mission
Barberini's Argument • Defeat in Rome
The Warning

In December 1615, a full year after the Dominican Tommaso Caccini had launched his attack from the pulpit and following a long bout of illness, Galileo finally set out for Rome with his secretary and valet to defend his name and to try to convince Church authorities in person that they were misguided in their opposition to Copernicanism. He had with him a draft of a paper he had been working on with increasing excitement, which he thought provided the necessary conclusive demonstration of Copernicanism. It was his theory of the tides, which showed how the daily rotation and yearly revolution of the Earth was the cause of tidal motion in the oceans. It was, of course, wrong, though Galileo seems not to have realized this until very late in his life, if ever.

It was a much different visit from his previous triumph of just four years earlier. A palpable chill was in the air where discussion of Copernicanism was concerned. The Tuscan ambassador to Rome, Piero Giucciardini, thought the whole enterprise ill-advised and irritably said so in official correspondence: "I do not know whether he has changed his theories and opinions, or his disposition, but this I know, that certain brothers of St. Dominic, who are in the

Holy Office [the Inquisition] and others are ill disposed toward him, and this is no fit place to argue about the Moon or, especially in these times, to try to bring in new ideas."[1]

No one could deny the brilliance of Galileo's discoveries, but now the matter of how they might best be fitted into existing patterns of knowledge and experience was at issue—and Galileo's position was seen as radical in the extreme. The Church, so recently savaged by the dogmatic dissenting opinions of those other radicals Calvin and Luther, was not eager to embrace the profound implications of the Copernican hypothesis. The Dominicans were expressing outright hostility and alarm; Galileism, it was whispered, was even more dangerous than Calvinism. The Jesuits, including those who had lauded Galileo just four years earlier, were quietly withdrawing from the field under orders from their general, who had commanded them to avoid supporting any position whatsoever that might weaken Aristotelianism.

Warning signals were coming from all directions. Galileo's friend Giovanni Ciampoli wrote with this description of a conversation he'd had with Cardinal Barberini (later Urban VIII) concerning the scientist and his opinions:

> Cardinal Barberini, who, as you know from experience, has always admired your worth, told me only yesterday evening that with respect to these opinions he would like greater caution in not going beyond the arguments used by Ptolemy and Copernicus, and finally in not exceeding the limitations of physics and mathematics. For to explain the Scriptures is claimed by theologians as their fields, and if new things are brought in, even by an admirable mind, not everyone has the dispassionate faculty of taking them just as they are said. One man amplifies, the next one alters, and what came from the author's own mouth becomes so transformed in spreading that he will no longer recognize it as his own. And I know what he means. Your opinions regarding the phenomena of light and shadow in the bright and dark spots of the moon create some analogy between the lunar globe and the earth; somebody

> expands on this and says that you place human inhabitants on the moon; the next fellow starts to dispute how these can be descended from Adam, or how they can have come off Noah's Ark; and many other extravagances you never dreamed of. Hence to declare frequently that one places oneself under the authority of those who have jurisdiction over the minds of men in their interpretation of Scripture is to remove this pretext for other people's malice. Perhaps you think I go too far in playing the sage with you; please forgive me, and thank the infinite esteem which makes me speak thus.[2]

Galileo complained that former friends and allies among the Church hierarchy were reluctant to see him in person, and he was reduced to sending them letters pleading his position. In a report to his patron in Florence, Cosimo de' Medici, he wrote:

> My business is far more difficult, and takes much longer owing to outward circumstances, than the nature of it would require; because I cannot communicate directly with those persons with whom I have to negotiate, partly to avoid doing injury to any of my friends, partly because they cannot communicate anything to me without running the risk of grave censure. And so I am compelled, with much pains and caution, to seek out third persons, who, without even knowing my object, may serve as mediators with the principals, so that I may have the opportunity of setting forth, incidentally as it were, and at their request, the particulars of my interests.[3]

Despite these difficulties, real or invented, he was not entirely silenced. He was, in fact, irrepressible. In a journal kept by a Monsignor Querengo there is an illuminating reference to the scientist:

> We have here Sig. Galileo, who, often, in gatherings of men of curious mind, bemuses many concerning the opinion of Copernicus that he holds for true. . . . He discourses often amid fif-

> teen or twenty guests who make hot assaults upon him, now in one house, now in another. But he is so well buttressed that he laughs them off; and although the novelty of his opinions leaves people unpersuaded, yet he convicts of vanity the greater part of the arguments with which his opponents try to overthrow him. Monday in particular, in the house of Federico Ghisilieri, he achieved wonderful feats; and what I liked most was that, before answering the opposing reasons, he amplified them and fortified them himself with new grounds which appeared invincible, so that, in demolishing them subsequently, he made his opponents look all the more ridiculous.[4]

Ambassador Giucciardini found this sort of behavior alarming and dangerous:

> He [Galileo] is all afire on his opinions, and puts great passion in them, and not enough strength and prudence in controlling it; so that the Roman climate is getting very dangerous for him, and especially in this century; for the present Pope [Paul V], who abhors the liberal arts and his kind of mind, cannot stand these novelties and subtleties; and everyone here tries to adjust his mind and his nature to that of their ruler. . . . Galileo has monks and others who hate him and persecute him, and, as I said, he is not at all in a good position for a place like this, and he might get himself and others into serious trouble. . . . To involve the Grand Ducal House in these embarrassments and risks, without serious motive, is an affair from which there can come no profit but only great damage. I do not see why it should be done, the more so when this happens only to satisfy Galileo. He is passionately involved in this quarrel, as if it were his own business, and he does not see and sense what it would comport [i.e., what its consequences might be]; so that he will be snared in it, and will get himself into danger, together with anyone who seconds him. . . . For he is vehement and is all fixed and impassioned in this affair, so that it is impossible, if you have him around, to escape from his hands. And this is a business which is not a joke but may become of great consequence, and this man is here under our protection and responsibility.[5]

And in a later communication with Florence the ambassador wrote: "Galileo has relied more on his own counsel than on that of his friends. The Lord Cardinal del Monte and myself, and also several cardinals from the Holy Office, had tried to persuade him to be quiet and not to go on irritating this issue. If he wanted to hold this Copernican opinion, he was told, let him hold it quietly and not spend so much effort in trying to have others share it. Everyone fears that his coming here may be very prejudicial and that, instead of justifying himself and succeeding, he may end up with an affront."[6]

One of those cardinals from the Holy Office who offered advice to Galileo was Maffeo Barberini. It was a meeting of great significance, because it was at this time that Barberini responded in detail to the root cause of the Church's discomfort with Galileo. The problem, he explained, was Galileo's view (often called scientific realism), implicit in his *Sunspot Letters* and explicit in the "Letter to Christina," that there is a single and unique explanation to natural phenomena, which may be understood through observation and reason, and which makes all other explanations wrong.

The counter-argument presented by Barberini, which he would continue to hold to throughout his papacy, was the argument from the omnipotence of God, and it is essentially unanswerable without stepping far outside the Christian tradition. We have it from Cardinal Oregius, who witnessed the conversation, that "when Galileo heard these words, he remained silent with all his science and thus showed that no less praiseworthy than the greatness of his mind was his pious disposition."[7]

Barberini's was a deceptively simple argument, one that René Descartes was to elaborate in his *Principles of Philosophy* exactly twenty-eight years later. Infinite power, it contends, is at bottom identical with the infinity of what is possible. This means that there can be no certainty about the cause of an event by examining its result. Any number of different paths could have led to the same destination. Thus, on the assumption of an infinitely powerful God, the experimental verification of hypotheses loses its power as conclusive demonstration.[8] Fortunately, as Descartes recognized, certainty as to the causes of events is not necessary for the conduct of life. What was important about a scientific hypothesis was not so much whether it was true as whether it worked.*

The crucial issue for Cardinal Barberini was not whether one scientific hypothesis could account for appearances better than another—whether Copernicanism more accurately described what had been observed in nature than Ptolemaism—but whether the successful hypothesis ought to be treated as truth. Could a scientific hypothesis, no matter how successful, be considered to be the uniquely correct description of nature, making all other hypotheses wrong? His answer, based on God's omnipotence, was no. Scientific hypotheses may be successful, but our reason will never be able to tell us whether they are true. Truth is a metaphysical issue—beyond physics—and it involves such questions as meaning and purpose, which are unquantifiable and therefore not amenable to scientific analysis.

Galileo, for all his genius, seems to have been unwilling or unable to grasp this point, which was central to the Church's attitude to the new science and its pretensions to access to truth. His failure to confront the argument in a serious way would be the central reason for his clash with the Inquisition seventeen years later.

The atmosphere in Rome went from bad to worse for Galileo: two months after his arrival in the city the committee of eleven consultants to the Holy Office that had been asked to report on Copernicanism announced the results of its deliberations. They could hardly have been less favorable.

The propositions the group had been asked to examine were the following:

- The Sun is the center of the world and hence immovable of local motion.

---

*It should perhaps be noted here that many modern scientific realists are what are technically known as *fallibilists*. Fallibilism proposes a reality that is "concrete" in the ordinary sense but at the same time contingent on possible redefinition in the light of new discoveries. It seems to me that fallibilists like to have their cake and eat it, since in their philosophy the point is never (or need never be) reached when a thing can definitively and nonprovisionally be called "real." In other words, fallibilist truth is always subject to change at a moment's notice. This is not what most of us think of as reality.

- The Earth is not the center of the world, or immovable, but moves according to the whole of itself, also with a diurnal motion.

The advisers found the first proposition to be "foolish and philosophically and formally heretical, inasmuch as it expressly contradicts the doctrine of the Holy Scripture in many passages, both in their literal meaning and according to the general interpretation of the Fathers and Doctors." The second proposition was declared to be deserving of "the same censure in philosophy and, as regards theological truth, to be at least erroneous in faith." Galileo was not mentioned.

To those in the know within the Church, this had been a foregone conclusion. In terms of Aristotelian physics, to which Dominicans and Jesuits alike subscribed, the movement of the Earth was indeed an absurd idea. This was true for a host of common-sense reasons that boiled down to the overriding fact that there seemed to be no sensory evidence to support the idea and much to deny it. And if the movement of the Earth was not a demonstrated truth, there was no need to question the existing interpretation of Scripture, which automatically put the proposition of the Earth's motion in the category of being contrary to Scripture and thus heretical.

Clearly, the climate was not right for Galileo or for Copernicanism. It was a period of extraordinary sensitivity regarding questions of scriptural interpretation and Church authority, and the Church's historically tolerant attitude toward science had been set aside in favor of what were seen as questions of survival. To the committee of consultants it would have seemed that there was little difference between the Protestant reformers' demand that individuals be allowed to interpret the Bible in their own way, and Galileo's insistence that Scripture be reinterpreted to be in accord with the (still highly speculative) Copernican hypothesis. It was a storm many at the highest levels believed would blow over, probably as a result of more complete confirmation of the movement of the Earth. But in the meantime it would be a wise man who would keep his head down.

In any case the advisers in their blanket condemnation seem to have gone too far even for the Holy Office. It issued no formal condemnation of Copernicus or Galileo. Instead, it took two steps that might be seen as conciliatory in the circumstances. The first was to ask the Vatican's senior theologian, Cardinal Bellarmine, to have a private talk with Galileo and order him to stop proselytizing his Copernican views. The exact content and circumstances of this warning are matters of great controversy among historians and were to have a grave impact at the time of Galileo's trial eighteen years later.

The second development arising out of the report of the consultants was a decree issued by the Congregation of the Index, the official Catholic board of censors. This was by no means as serious as a ruling by the Holy Office, but it was nonetheless significant. The Index decree said four things: it repeated the advisers' contention that the doctrine of the Earth's motion was false and contrary to the Bible and a danger to the Catholic Church; it condemned and prohibited a recently published book that attempted to reconcile Copernican views with Scripture* (but took no action against its author); it suspended circulation of Copernicus' own book, pending revisions; and it provided for similar action to be taken with regard to other books on the same subjects. The revisions to Copernicus were completed in 1620 and amounted to minor changes in about a dozen passages that either dealt with religion or implied that his description of planetary motion was meant to be taken as a portrayal of physical reality rather than a mathematical convenience.

The draft document prepared for the Cardinals of the Congregation of the Index and presented to them April 2, 1618, is seldom included in collections of material related to the trial of Galileo, but it ought to be. It was submitted to the mathematicians of the Collegio Romano for their approval before being used as the basis for the decree of 1620 announcing the revisions to *De Revolutionibus*. It urged a peaceful coexistence between astronomical science and theological doctrine, by means of a *via media* or middle way that would allow apparent contradictions between the two to be accommodated. The proper method of astronomy, the document said, was the use of "false and imaginary principles in order to save celes-

---

*The book was *Letter of the Rev. Father Paolo Antonio Foscarini, Carmelite, on the Opinion of the Pythagoreans and of Copernicus concerning the Motion of the Earth, and the Stability of the Sun, and the New Pythagorean System of the World.*

tial appearances and phenomena," adding that "it is customary for the science of Astronomy in particular to make use of false suppositions."[9] This approach allowed the more contentious passages of Copernicus to be interpreted as breaches of scientific etiquette rather than heresy. This was a position that had been under active development within the Collegio Romano throughout the early seventeenth century, and of course it is reflected in the stance taken by Cardinal Bellarmine, the head of the college.

A consequent subtlety of the decree that often goes unnoticed is that it makes a distinction between scientific hypothesis and theological interpretation. Thus, the book that was condemned and prohibited dealt with religious interpretation, a field the Church insisted must be left in the hands of the Fathers and the Doctors —that is, the theologians of time-tested ability, inspiration, and insight such as Augustine and Aquinas. Copernicus' book, in contrast, was seen as mainly scientific hypothesis and was merely withdrawn pending deletion of theological references.

There is strong evidence that Pope Paul V had wanted to declare Copernicus contrary to the faith and have his book banned but was persuaded to accept milder action. A contemporary diary states that "Paul V was of opinion to declare Copernicus contrary to the faith; but the Cardinals Caetani and Maffeo Barberini withstood the pope openly and checked him with the good reasons they gave."[10] This is substantiated by a much later aside in a letter written by Cardinal Barberini, who had by then become pope, referring to "these difficulties of which we relieved Galileo when we were Cardinal."

Galileo apparently accepted with resigned good grace Bellarmine's injunction to stop his promoting of Copernicus. As the two were old friends, despite their differences over cosmology, their private meeting appears to have been cordial. Neither the Index edict nor his personal warning, it seemed to the scientist, prevented him from continuing to amass scientific evidence for the Copernican theory, as long as he refrained from arguing his position in public. In time, and with better evidence, he was able to assure himself, the Copernican position was bound to carry the day. In a forty-five-minute audience with Pope Paul V prior to his departure for Florence, he was told that there was no thought of personal sanctions against him.

There was one continuing irritation. Embarrassing rumors persisted in Pisa and Florence that he had secretly been hauled before the Inquisition and made to recant his Copernican views. To set the record straight, Galileo asked for, and was given, a letter from Cardinal Bellarmine stating that the rumors were false and describing what had in fact taken place. The contents of the letter would play an important part in Galileo's trial:

> We, Roberto Cardinal Bellarmine, having heard that it is calumniously reported that Signor Galileo Galilei has in our hand abjured and has also been punished with salutary penance, and being requested to state the truth as to this, declare that the said Signor Galileo has not abjured, either in our hand, or the hand of any other person here in Rome, or anywhere else, so far as we know, any opinion or doctrine held by him; neither has any salutary penance been imposed on him; but only the declaration made by the Holy Father and published by the Sacred Congregation of the Index has been notified to him, wherein it is set forth that the doctrine attributed to Copernicus, that the Earth moves around the Sun and that the Sun is stationary in the center of the world and does not move from east to west, is contrary to the Holy Scriptures and therefore cannot be defended or held. In witness whereof we have written and subscribed these presents with our hand this twenty-sixth day of May, 1616.[11]

Galileo carefully stored the document with his most important papers. It was, it seemed to him, a passport that would in due time provide safe passage across hitherto forbidden frontiers of expression.

FOURTEEN

A Dialogue in Venice • Science's Successes
The Nature of Knowledge

I had, without much difficulty, persuaded Berkowitz to meet me in Venice. I wanted to introduce him to Sister Celeste so that the three of us could have the pleasure of discussing Galileo in the very place where he had set his famous *Dialogue on the Two Chief World Systems.*

The meeting of our little debating society took place in the Danieli Hotel. I spotted Berkowitz and Sister Celeste in conversation.

"I see you've already introduced yourselves," I said.

"Yes, and Mr. Berkowitz kindly bought me an aperitif."

She saluted him with her glass and polished off what remained of her drink.

"She wasn't hard to find in this crowd," he said, gesturing to the well-heeled vacationers populating the room.

We adjourned to the rooftop terrace bar. The evening sun was painting in reds and oranges the Baroque marble frippery of the imposing church of Santa Maria della Salute across the lagoon. In the foreground we could see the temple facade of that sixteenth-century masterpiece of Pythagorean harmony and proportion, the church and monastery of San Giorgio, and beyond it the dome and spires of Il Redonte, the Church of the Redeemer, both designed by the architect Andrea Palladio. The sun burnished the gleaming mahogany planking and bright brass fixtures of the scurrying water taxis on the broad Bacino di San Marco. The Lido, with its fabled Edwardian resort hotels, glittered like a necklace of hammered copper on the horizon. Behind us, the Dolomites loomed a smoky azure through the summer haze. We were moved to silence by the beauty surrounding us.

I finally broke the spell.

"On this scale, with four- or five-story buildings and the occasional tall spire and the narrow streets and the way they wind about rather than following a grid pattern and so on, the works of man don't seem to offend against nature, do they? They seem in a way complementary. We seem to fit into the scheme of things, if you know what I mean."

"I'd go further than that," Berkowitz said. "The way the city brings the sea right into the streets and is so completely integrated into its natural environment, you have to think it's an improvement on nature. I mean, this place, taken as a whole, is surely a work of beauty that would challenge anything in the so-called natural world. Think about it. The natural landscapes that we so admire are shaped by the life that occupies them—the animals and the plants. What we have here is a case of the animal called man having shaped the natural environment in a way that is undeniably beautiful and successful, don't you think?"

"That is because this city was built in the time before men began to think of themselves as gods," Sister Celeste observed, still staring dreamily out over the water. She gave her head a little shake and sat up straight in her chair, as if she'd been admonished by a teacher, before continuing. "It is also the city where Galileo did much of his engineering work, in the Arsenale shipyards very close by here." She turned to point to her right. "And there is the campanile of San Marco, where he first demonstrated his telescope to the Doge Leonardo Donà in 1609. Of course it is not the same tower but a copy. The original was built in the twelfth century to be a lighthouse. It collapsed in 1902, when its foundation gave way, and had to be rebuilt."

Berkowitz, only half listening, had been poring over his *Eyewitness* guidebook to Venice, in which it seemed he had made a discovery. "You can say what you want about the hubris of modern scientific man," he said, closing the book and keeping his place with his index finger, "but it says here that two of those churches over there, the Redonte and Santa Maria della Salute, were built to give thanks for the ending of epidemics of plague. That's something we don't have to deal with nowadays, thanks to science."

"Of course you are right," said Sister Celeste. "And it is interesting, do you not think, that when, with the help of science, we eradicated smallpox only a few years ago—as deadly a plague as ever there was—not a single church was built anywhere in thanksgiving. Not to my knowledge. This was a disease that all but wiped out the aboriginal population of the Americas, to say nothing of the devastation and disfigurement it caused among the other peoples of the world. We give thanks to no one but the scientists that it is gone. I would certainly feel better if we had built a cathedral, or even a humble church. Even Bertrand Russell, who has said that medieval man was too humble before God and whom you might guess is not among my favorite philosophers—even he was afraid of a certain intoxication of power that he called a madness in modern man. He had concluded that a degree of humility in the face of the unknown is necessary for sanity. He said that any philosophy that contributes to the intoxication of power, the blindness to our limitations, pushes us all a step closer to a vast social disaster. He mentioned especially the Pragmatist philosophers, for whom knowledge and values are simply tools for organizing experience in instrumentally useful ways. What we think we know in any deeper way is fantasy and nonsense. It is, I would venture to say, if not the most widespread then certainly the dominant philosophy of our time."[1]

Berkowitz shifted uncomfortably in his chair but said nothing. I knew he was sympathetic to the Pragmatist view. Then I remembered the last line in the paper Sister Celeste had lent me in Ferrara. "You said an interesting thing, Sister, at the end of your notes on Aristotle and Aquinas. You said, I think, that in the modern world what chiefly distinguishes science from religion is the matter of success. What did you mean by that? I would have thought that what distinguished them was their subject matter, and the question of faith versus fact."

"And it was indeed the question of faith versus fact that I had in mind. Let me see if I can explain in only a few words." She paused a second, then laughed. "But of course I cannot explain in a few words. If you want to know my opinion, you will have to endure the whole argument, which may take some time."

"Please go ahead," I urged her. "What could be finer than considering matters such as these in so beautiful a setting."

Berkowitz nodded his enthusiastic assent as well.

"Well, then, we might as well begin with the very basic idea that both religion and science are concerned with the acquisition and verification of knowledge about the universe. Why we want that knowledge is another question. Aristotle said that man is defined by his curiosity, that man is uniquely the animal that seeks out knowledge for its own sake. Others say that we seek knowledge of the world in order that we might live in it with comfort and security. Our curiosity, in other words, is a survival mechanism. We could even include spiritual knowledge in this viewpoint, if we say that we seek spiritual knowledge because it helps us to survive in some way."

Berkowitz held up a finger, to interject. "Sister, with all due respect, that is a tautology, a circular argument. You are begging the question, by positing first that we seek knowledge in order to survive, then pointing out that we seek spiritual knowledge along with other kinds, and then concluding that spiritual knowledge must play a role in survival. That, I'm afraid, is not a logically sound defense of the value of spiritual knowledge. As you know, Aristotle's *Posterior Analytics* says there is no necessity in any conclusion from an argument that uses its own conclusion as its starting point. The argument 'If all cats are black, and this is a cat, then it must be black' doesn't prove the cat is black, unless you buy the premise that all cats are black as an article of faith, or you have separate proof."

Sister Celeste nodded and smiled at him. "But then you will have to agree that the same logical flaw applies to the survivalist argument for seeking any other kind of knowledge, as well. Am I not right? We say, for example, that we seek knowledge in order to survive; that it may be observed by anyone that we seek a specific kind of knowledge we call 'scientific'; therefore, it is concluded that scientific knowledge helps us to survive. The very same circularity is present, is it not?"

Berkowitz, frowning, leaned forward and massaged the space between his eyebrows. "Mm," he said.

"So I would put that rather fruitless issue to one side and simply agree with Aristotle that curiosity is what defines us, and leave it at that. What is important to our discussion here is not why we seek knowledge, but how we decide what is knowledge and what is not—as one might say, what is fact and what is fancy."

"In other words, what we should be concerned with is techniques for knowledge verification," I suggested.

"Yes, I think so. And I would say that science has been such an enormous success since Galileo's time because of its success in knowledge verification. It not only has knowledge, it has techniques for demonstrating the veracity of that knowledge."

"Whereas religion does not have means for verification?" Berkowitz suggested.

The little nun shook her head. "I would say rather that religion has essentially the *exact same* methods for verification, and that in the present time there is a strong cultural bias against using them. Religious truths go unacknowledged and unaccepted not for any reasons of verification, but because our culture is willfully blind to them. We are blinded by the material successes of science."

Berkowitz smiled and shook his head.

"Yes," Sister Celeste insisted, "willfully blind. We all, I think, understand that there is a spiritual side to life, and yet we, most of us, decline to admit the validity of spiritual knowledge in the conduct of our daily lives. And the reason is simply that we have been instructed not to do so. Let me explain.

"Many ways have been suggested over the ages for verifying knowledge of the world and the ways it works. But let us take as an example the method proposed by the Pragmatic philosophers such as Charles Peirce and William James and John Dewey. Theirs is essentially the scientific view. It says that secure knowledge is gained by sensory experience. What our senses tell us about a thing is what it really is. So in order to know something, we must be able to experience it. This is empirical knowledge—*empeira*, the Greek for experience. Pragmatists and scientists would say this is the only kind of reliable knowledge, and the only kind worth paying any heed.

"If you think about it for a moment, however, you will realize that all sensory experience is subjective. Only I know what the color blue looks like to me, or how water tastes, or how heavy a brick feels in my hand. Subjective knowledge is of use only to me. It is objective knowledge that civilizations are built upon—a body of objective knowledge that everybody agrees with, and that may be added to from time to time and modified with new learning.

"And how do societies turn our subjective knowledge as individuals into a body of objective knowledge accessible to everyone? They do it by exchanging experiences, talking things over, and coming to a consensus of opinion. We agree that this bit of the electromagnetic spectrum will be called 'blue'; that this experience of resistance will be called a weight of such-and-such. Empirical knowledge—and this includes scientific knowledge—is rooted in consensus. The broader that consensus, the more we feel justified in relying on the knowledge. I, for one, would be reluctant to go aloft in an airplane built according to a theory of aerodynamics that was rejected by the Society of Aeronautical Engineers. Yes?"

"Most emphatically yes," said Berkowitz.

"But as any scientist will tell you, the process of gaining empirical knowledge is not simply a passive one of receiving the input from our senses. We continually verify what our senses are saying by experiment—in other words by acting on that information to test it."

"Is this a dagger which I see before me?" Berkowitz intoned in a deep baritone. He snatched at the thin air over the table.

Sister Maria Celeste smiled. "You are an astute listener, Mr. Berkowitz."

"But still," I said, "in the end it comes down to consensus even for science, doesn't it?"

"Indeed it does, for what else is there to base knowledge on?"

I waited for her to answer what seemed like a rhetorical question, but she merely sipped her drink.

"That's an odd question for you to ask, Sister," I said finally. "What about the knowledge provided by revelation, by spiritual experiences?"

Berkowitz sniffed his disdain.

"Be careful, Mr. Berkowitz, before you too hastily reject what we may call religious knowledge. Why would you dismiss it out of hand?"

"Because I have no proof of its assertions, and you have none to offer me. I believe what I can verify with my senses. I believe this glass is here before me because I can see it and touch it. I believe that if I drop it, it will hit the floor."

"No need to demonstrate, Mr. Berkowitz; I accept your point. But you believe a great deal more than that, do you not? For instance, do you not believe that matter is made up of a host of subatomic particles? And yet I do not think you have actually seen one of these."

"Yes I do believe that, and the reason is that other people, scientists, *have* seen them." The little nun began to interrupt, and he quickly backtracked. "I mean to say, they haven't actually seen the actual *thing*, if you know what I mean—the quark or whatever —but they have seen clear evidence of their existence, in cloud chambers and on photographic plates and in test instruments and what-not."

"I think I see where you're going with this, Sister," I said. "The proof science has to offer our friend here of, say, the existence of quarks or mesons, is a consensus of scientific opinion based on experiments in which the actual object is not observed but only its effects on some environment like a cloud chamber. So his belief in the existence of the constituent elements of matter derives from the acceptance of the authority of a group of people who claim to have seen solid evidence of subatomic phenomena. He has no firsthand evidence, nor can he have, unless he himself becomes a scientist and is, as it were, initiated into the profession."

"Yes, and at another level I would point out that the whole idea of causation in nature, on which science relies so heavily, is completely unverifiable. One can never observe a cause-and-effect process in action, can one? All that is observable is that one phenomenon follows after the other. The rest is conjecture, or faith. And I would add that in order to be initiated into the society of scientific experts—that is, to pass exams and become accepted by his or her peers and elected to professional societies and be given the opportunity to do useful work—the prospective scientist must accept as true a vast body of existing knowledge. The candidate commences work on that foundation of preexisting knowledge, which must be accepted on faith. It is as St. Augustine said: 'Seek not to understand so that you may believe; but believe in order that you may understand.' People sometimes misunderstand this as meaning you must sacrifice knowledge to faith. But it really means that faith is offered to reason as the disclosure of the possibility of reason's self-fulfillment. Faith is the point of departure of knowledge."

"So really, Sister, what you are saying is that there is essentially no difference between scientific and religious knowledge, if we're referring to the way we verify knowledge in order to decide how reliable it is."

"I am saying that both forms of knowledge are based first on subjective experience, then on consensus, and finally acceptance of authority. Let us say you happen on a person on her knees in a church. She appears to be speaking to someone, perhaps only silently. You ask her, 'What are you doing?' and she says, 'I am talking to God.' You ask, 'But how do you know He is listening?' and she replies, 'Because I have had experiences in the past in which my prayers have been answered in various, often surprising and unexpected, ways. Moreover, the Holy Scriptures tell me He is listening, and so does the Church and all its greatest thinkers over the past two thousand years.'

"Now, if you are a scientist, and if you take no time to reflect, you may say to yourself that this is no proof at all. But I think I have demonstrated that it is indeed proof of a nature very similar to, if not identical with, the proof that undergirds all science. That is to say, empirical, experiential evidence, and the accumulated wisdom of authority."

Berkowitz was becoming agitated, as he did when his skepticism was aroused. It seemed a propitious moment to adjourn our discussion, for the time being.

"On that note," I said, "I suggest that we retire to the dining room. I'm told the seafood is superb."

By now the lights on the lagoon islands were twinkling magically against an indigo sky, while their reflections danced in the lagoon.

"Only on condition that we continue this tomorrow," said Berkowitz.

"That's the plan, is it not, Sister? At lunchtime?"

We agreed on a time and place.

FIFTEEN

Further Dialogue in Venice

Science and Faith • Faith and Reality

The Nature of Progress

The following day, I found my two friends waiting for me on the narrow terrace of a small café alongside the San Provolo canal. Gondolas plied up and down, and directly across from us was a beautiful pink palace in the Oriental-looking Venetian Gothic fashion that had preceded the Renaissance style.

I had interrupted them in mid-conversation, and they carried on as if I'd been there from the start.

"But, Sister," Berkowitz was saying, "you yourself have already pointed to the big difference between the way science and religion operate. Science is continually verifying its knowledge by direct experience and through experiments. We accept scientific knowledge because it *works!*"

"But, Mr. Berkowitz, religious knowledge works, as well," the little nun responded. "That is why there are so many of us religious persons in the world." She laughed briefly, her eyes sparkling. "Its objective is different, but it works, just like science does. I can tell you for a certainty that there is very profound knowledge of the universe to be had through religious experience. Very profound indeed. Nothing I have read of science has so convinced me, or so changed my life. As for the so-called scientific method, that is a separate and very interesting issue worthy of a discussion of its own. For the moment, I would prefer to stay with our original point, with your permission, of course.

"You must understand that often in order to gain knowledge it is necessary to approach any subject with a degree of belief, or at least an absence of skepticism—a willing suspension of disbelief, as they say in the theater. If I am to learn from a teacher I must believe that what she tells me is true. If scientists did not accept on faith that the universe is basically orderly and amenable to organized study, and that mathematics is a reliable tool, they would never bother to pursue their studies, because it would seem pointless. What except faith prompts a scientist to believe in what is unobservable, like the linkage of cause and effect, or the existence of infinity, or necessity? In the same way, if one is to gain religious knowledge, it is necessary to approach the subject with a willingness to believe. Without an open mind, Mr. Berkowitz, I am afraid one is unlikely to learn anything."

The comment was meant in a kindly way, but I could see that Berkowitz was stung by it. Sister Celeste was quick to realize the unintended effect of her remark. It was an uncomfortable moment.

Finally I said to Sister Celeste, "There is a moving expression of what you're saying in William James's essay 'The Will to Believe.'"

"Ah yes," she said, "such a fine, lovely man, William James. He was a Pragmatist, Mr. Berkowitz, much like you, I think."

Berkowitz smiled at this artless attempt to mollify him.

"Perhaps you are not familiar with the essay?" she asked.

He shook his head.

"James begins by saying that there are some truths that become obvious only when approached with a preliminary faith in their existence. For instance, it is necessary to have a preexisting faith in the reality of human solidarity in the face of adversity—you could say, of charity—in order for the fact of its existence to be established. He uses the example of a train being robbed by bandits, not an unusual occurrence in his day. If, among the individuals on the train, there is a faith in the likelihood of fellow passengers coming to one another's aid, the bandits, who after all are greatly outnumbered, will be overcome by the collective action of the passengers.

Solidarity will have been shown to exist. If, on the other hand, such a faith does not exist, then no single individual will risk taking on the bandits and the robbery will succeed. Without belief, in this case, the object of the belief does not exist."

"I can think of better examples than that," I said. "How about the international currency system? It used to be that all paper money was redeemable at the bank in specie, usually gold. Now, of course, there is nothing backing currencies except faith. Faith in things like the basic solidity of an economy, for example. Faith that when you go to the store tomorrow, the merchant will accept it in exchange for the goods he sells. The paper itself has no intrinsic value. So without faith, money would not work. Now that I think of it, no one would ever drive a car if they didn't have faith that other people on the road, by and large, would follow the rules responsibly."

"Come now," Berkowitz said. "All you are saying is that people operate on the basis of well-founded expectations, or well-justified beliefs. In other words, they decide what to do according to beliefs that grow out of evidence. That's obvious. What's that got to do with faith?"

"I'll tell you what it has to do with faith," I said. "If you think of the examples we've been using—train robberies, money, traffic, and so on—it's impossible to separate faith from the so-called objective evidence."

"Huh? I don't get it."

"What I mean to say is that the 'evidence' you speak of would not exist if faith had not created it in the first place. To some greater or lesser degree—exactly how much is probably impossible to know—but to some degree the evidence depends on faith having created it. Your so-called objective reality is to some extent a creation of faith. Now that I stop to think about it, I can see all sorts of ways that faith operates in daily life. Economies could not function without it. Nor could human relations. In fact, faith is a very common commodity, isn't it? Religious faith is just one variant of a very widespread phenomenon."

"All right, I get the picture," Berkowitz interjected somewhat impatiently. "But we've gotten way off track. I still would like to know how you explain the fact that scientific belief has come to be dominant, when just 350 years ago religious thought was so utterly in control. How do you explain science's success if it's not a superior tool?"

I suggested the obvious: "We shouldn't assume that the ascendancy of the scientific worldview is in any way permanent or immutable. After all, 350 years is not much compared with the two thousand years of prior religious dominance. It could be just an aberration, a blip on the screen."

"Some blip," Berkowitz snorted.

"I'm serious. There is a good argument for the historical contingency of scientific culture. You can't assume that its success was inevitable, or that it's here to stay.[1] The argument goes like this: The Age of Faith ended when people were no longer able to believe in the idea of Nature as Providence, as existing solely for the welfare of humankind. Before the Scientific Revolution of the seventeenth century—Galileo and all that—for most of recorded history people did not see their condition in the world as being defined by fundamental want or physical need, as we do. They saw themselves as the beneficiaries of an all-providing Providence. The role played by technology was to assist and supplement nature, not to conquer and subdue it. For many centuries these tools and implements hardly changed at all. You could count the big technical innovations of a thousand years on the fingers of one hand: water power, wind power, the ox yoke, and the stirrup. Those wants and needs that were recognized and acknowledged were explained as resulting from humankind's failure to distribute the bounty of Providence properly. That of course put the question of justice front and center in their discourse.

"With the loss of faith in Providence that coincided with the Scientific Revolution and the Enlightenment, about the end of the seventeenth century, humanity was put in the psychological position of having to fend for itself in an essentially hostile world. We had to subdue and conquer nature in order to survive. That is what led to the sudden and massive upswing in scientific endeavor: it was not simply a continuation or acceleration of an existing trend. It was a new phenomenon that resulted from a new consciousness. And that new consciousness came out of the scientific idea of an uncaring universe.

"What all this means is that our present scientific culture is based on an idea that is both recent and controversial—the idea of the uncaring universe. If people were for some reason to lose their faith in that idea and adopt another, scientific culture as we know it would be transmuted into something else. It could happen—in fact, given time, it is *likely* to happen.

"And by the way," I concluded, "it's also true that the idea of a beneficent nature was a serious threat to science, because it weakened the rationale for scientific inquiry. There is no real need for scientific inquiry if nature is all-providing and basically reliable. That, it seems to me, is part of the reason why science has been so violent in its attacks on the religious idea of a teleology in nature. The idea of nature existing to serve an ultimate good, or more specifically human welfare, undermines the requirement for scientific exploration."

"That's all very interesting," said Berkowitz, "but increased effort in science needn't necessarily have led to increased success, if you know what I mean. I still would like you to explain why science succeeds so brilliantly."

Sister Celeste raised a tiny finger. "A fair question, Mr. Berkowitz, and it has an answer, I believe. First of all, as I have said, we—you and I, the family in that gondola—are constantly in the process of verifying our knowledge of the world by acting on it and in this way testing it. This is how we deepen belief in what we know. If society discourages us from acting on religious knowledge, it also limits the extent and depth of belief. So in a society such as ours that marginalizes religious experience and frowns on the too-public demonstration of religious belief, the growth and expansion of religious truth is stunted. Relative to science, then, religion will be less successful.

"Second, in the modern world religion and science focus their main interests in two different areas: religion on the spiritual side of things and science on the material. Any progress made by science is very noticeable, because it affects our physical circumstances in one way or another. Its successes are very tangible. The changes it causes are palpable, and for that reason it seems very powerful.

"More than that, though, virtually any and all scientific activity is, by the rules of the process, guaranteed to provide some form of 'success'—both the verification and the discrediting of hypotheses are treated as successes. Furthermore, even such clearly undesirable developments as the invention of weapons of mass destruction are also considered to be scientific successes. We say science is making progress as long as it is accumulating more facts and information. Since this is presumably an infinite process, the goal, the scientific millennium, recedes into the infinite distance. That is very convenient in a way, because it means that it never has to be examined very closely or defined precisely.

"But genuine progress really requires a concrete goal, do you not think, Mr. Berkowitz? Ask yourself, in terms of your personal life, 'How do I tell if a change is progress or not?' and I think you will discover that you cannot tell unless you have a concrete goal in mind. Change is merely change without a goal or object. Only change that brings one closer to a defined goal is progress."

"So you're saying that since science has no goal beyond more of the same, more facts of all kinds, it's unreasonable to call its trajectory 'progress'?"

"I would say so, yes. I would also say that its use of the term *success* is very problematic in the absence of well-defined goals. Success and progress are both virtually meaningless terms in the context of the broad, historic scientific endeavor, because of this lack of a teleology."

"What about discovering the truth about the world? That, surely, is a goal," Berkowitz said.

"It is, but it is not a goal of interest to science, I believe. The 'truth about the world,' as you put it, is in the end a metaphysical concept. Science is interested not in the truth but in the facts about the world."

Berkowitz was ready with a response. "Okay, for the sake of argument I'll grant you all that. But so far, Sister, you've only succeeded in pointing out why religious knowledge shouldn't be denied the possibility of success. You've failed to show that it *does* have successes in any way comparable to those of science." He folded his arms and looked first at me, then at her, shrugging.

"Let me then suggest a situation in which such success might be more obvious to everyone," Sister Celeste said. "Because I think the problem is one of perception rather than performance. Remember that the trajectory of science cannot properly be called success because it has no goal, no teleology. Religion, on the other hand, does have such a goal. Imagine for a moment what it would be like if we lived in a religious rather than a materialist society. Then people would have a common idea of what constituted advance and retreat and could tally up progress from time to time.

"But beyond that I would say that in cases where faith—the 'will to believe'—can help establish a significant fact, it would be an insane logic that would refuse to accept that fact as trustworthy because it is 'unscientific.' Goodness may not be quantifiable, but it surely exists. Everyone, I think, has some degree of belief in the existence of a realm of spirituality that is not accessible to scientific examination but is nevertheless real. It seems to me to be *pazzo* [here she tapped her temple with a finger] to discourage the 'willing advances' Mr. James suggests are required of us if we are to tap into that very rich realm of knowledge, just because this is 'unscientific' behavior. It may be unscientific, but it is nonetheless a valid approach to the gaining of knowledge. And as I said a few moments ago, it is essentially the same approach that science uses at its deepest and most fundamental levels."

Berkowitz chuckled. "Very good! So a decision not to pursue religious or spiritual knowledge is an irrational decision—one could even say an 'unscientific' decision?"

"Yes, I believe so."

Berkowitz seemed flushed. "Well, I must congratulate you. That was very nicely put, Sister. I'm not sure I'm a hundred percent convinced, but I give you full marks for rhetorical elegance."

"Yes indeed," I said, raising my glass.

"*Mille grazie*, gentlemen, you are too kind," Sister Celeste said. Her coy smile belied an honest delight.

"And now I am afraid I must leave. Family duties call."

SIXTEEN

On the Lido • The Problem of Objectivity

Reasoning's Recursiveness

The Map and the Territory

An Unfortunate Outburst

Early the next morning, the telephone rang. It was Berkowitz, in a highly agitated state.

"We have to get together."

"Who?"

"The three of us. Today. Right away."

"Why today? I thought we were going to meet back in Rome."

"It won't wait that long. She pulled a fast one on me yesterday, my friend. I woke up at about three this morning and it was perfectly clear to me. In all the talk about truth verification she left out the most important tool science has. She said she didn't want to talk about the bloody scientific method, for God's sake. And I let her get away with it! The scientific method is what it's all about. I could kick myself. I have to talk to her today or I'll go nuts playing back might-have-been conversations in my head."

"As I remember," I told him, "she didn't refuse to talk about it, she just said it was a whole separate topic."

"Whatever. I got the hotel operator to find her number on the Lido, and I'm going to give her a call and set something up for this morning. Feel free to join us, or not."

"I wouldn't miss it," I said. I wondered briefly how he'd gotten Sister Celeste's family name, and then remembered she'd given him one of her calling cards. We knew she was staying at the home of her parents on the Lido.

I had just stepped into the shower when he called back to tell me we'd be meeting in the breakfast room of the Hôtel des Bains in an hour.

"We can't meet at the house because her parents aren't up yet."

"I'm not surprised. It's 8:00 A.M., and it's Saturday. I should be in bed, too."

"Stop whining. I'll meet you at the vaporetto stop in front of the hotel in fifteen."

Half an hour later we were stepping off the number 6 vaporetto onto the steel docking ramp at the foot of the broad, tree-lined boulevard called Santa Maria Elisabetta. It cuts through the Lido about a quarter way along the island's seven-mile length.

"Should we get a cab?" Berkowitz asked.

"It's only a five- or ten-minute walk," I said. "We might as well save some money."

The walk across the island from the lagoon side to the oceanfront was a pleasant one in the relative coolness of the morning. The air was soft, and the famous clarity of the Venetian light made our eyes feel young. A cluster of shops and restaurants around the little square where the vaporetto stops soon gave way to pleasant stuccoed villas and small apartment blocks hiding behind wrought-iron fences and tall hedges and shaded by enormous willows. Looking down paved cross streets we could see more of these sedately stylish homes, built as vacation retreats a hundred or more years ago when the Lido became Europe's most fashionable seaside resort. They were now mainly owned or rented by Venetians, who commuted to work in the city each day. Private motor launches lined the little canals that thread the island. The few cars that reach the island by ferry are far outnumbered by bicycles. A quiet, park-like retreat on the doorstep of one of the wonders of the world, the Lido seemed an agreeable place to call home. For me, it did much to explain Sister Celeste's quiet urbanity.

We soon found ourselves with the Adriatic at our feet, and turned right to follow the oceanside drive named for Marconi. A few moments later we strode through a gap in the tall hedge that shields the island's most glamorous hotel from prying eyes. We could see, a half-mile up the road, the fantastic turrets and phony crenellations of that other Lido landmark, the Grand Excelsior. Built in the era of the *Titanic* and proudly advertised as the biggest resort hotel in the world, it is the architectural prototype for goofy Las Vegas extravagance. Hôtel des Bains is another cup of tea, altogether more elegant and subdued, favoring polished hardwoods and brass over the Excelsior's marble and gilt. This was the setting for Thomas Mann's slightly creepy *Death in Venice*, both the 1912 novel and Luchino Visconti's 1971 film. It is a Dirk Bogarde–Gustav Mahler kind of place.

"He's dead, you know," Berkowitz said as we mounted the front stairs.

"Who's dead?"

"Dirk Bogarde. Died just a little while ago."

We crossed a vast creaking, gleaming hardwood floor in the main lobby to the lounge, where coffee and rolls were being served to a handful of hotel patrons. Sister Celeste had already taken a table, and as we pulled up our chairs a young waiter asked, *"Caffè?"*

"Si!" The three of us spoke in unison, and the little nun laughed, and added *"grazie."*

Berkowitz got down to business: "It's good of you to do this, Sister."

"Not at all. You seemed rather distraught on the telephone. But I have to warn you I am a little pressed for time this morning."

"Then I'll get right to the point. When we spoke yesterday you argued—very convincingly, I might add—for the equivalence of the religious and scientific ways of verifying truth, of defining reality."

"Yes. I said that as far as I can see there is nothing in the scientific approach that would allow it to claim a greater certainty or security for its knowledge."

"Exactly. But you deliberately excluded all reference to the scientific method, which is the secret formula, if you will, by which science establishes truths that are beyond doubt because they have been experimentally confirmed. It was not a fair comparison."

"Well, Mr. Berkowitz, I also refrained from bringing up the question of knowledge gained by revelation, though I could have argued that it, too, is empirical knowledge. So the comparison was not perhaps as unfair as it seems to you. In any case, let me see if I can satisfy you in the short time we have this morning."

Her back perfectly erect, she placed her wrists on the table and folded her fingers together.

"There are two basic arguments against the scientific method's usefulness in epistemology, in establishing what is real and true. The first is that scientists themselves do not really operate that way, and the second is that it is not as objective or value-free as it is held up to be."

"I'm not entirely sure what is meant by the scientific method," I said. "As I understand it, it's a blend of empiricism and experimental verification. A scientist makes observations of things happening in nature, and then he goes home and thinks about them and comes up with a theory to explain why they happen the way they do—"

"That's right," Berkowitz interrupted. "Or sometimes he may have the idea first and then make observations. In either case, he devises experiments to test his hypothesis, and they must be experiments that can be duplicated by other scientists so that they can test his theory independently. If the hypothesis stands up to experimental testing, it is accepted as fact. If the experiments don't confirm it, it is either modified and tested again, or discarded. Do you accept that definition, Sister?"

"It will do for our purposes," she replied. "Since you mention theory, let me start with the second of the difficulties I referred to. The question of the objectivity of experiments, or the lack of it, is usually talked about in terms of the lack of a clear separation between theory and experiment. As you know, since experimental physics deals with subatomic entities so unimaginably tiny that we humans are as big in relation to them as the known universe is to us, observations must be made with instruments that measure their impact on their surroundings. We can never see them, even with the most powerful of microscopes. So when a physicist sets up an experiment to test a hypothesis or theory, he receives the results of the experiment in terms of the readouts of various electronic detectors, which are probably linked and mediated by computers. Yes?"

Berkowitz nodded.

"This data, of course, is not like simple sensory input, like touching and seeing. It must be interpreted in order to make any sense. Yes?"

Berkowitz nodded again, more slowly.

"How do we interpret data of any kind? We do it by trying to fit it within the framework of what we already know, like a piece of a jigsaw puzzle. There is no other way. So how do physicists interpret the results of their experiments? By fitting them within the framework of current theoretical ideas! You can see, I think, Mr. Berkowitz, that theory and experiment are inextricably tied up together in a kind of recursive loop. There is a built-in perspective, we might even say bias, in the process of interpretation. Experiments do not test the world so much as such-and-such a theory about the world. Experimental evidence, to the degree that it must be interpreted, is not *objective* evidence. It always incorporates the point of view of the interpreter. Do you follow me?"

Berkowitz was frowning. "That may be a very nice point in logic, but I don't see how it has any application in the real world. The fact is we actually *do* learn a whole lot of very useful things via the scientific method."

"Yes indeed, we do," Sister Celeste replied. "And that is a puzzle, isn't it? How is it that science provides us with information we can use in our day-to-day life in this world, if its basic method is in this way circular, or we might say, self-referential? How does it reach out beyond the bubble of its own theory and touch the real world?"

I was beginning to get excited by all this. "You've reminded me of something that's always bothered me since I first read it. Charles Peirce has the most elegant definition of human reason I've seen, and it goes like this: 'The object of reasoning is to find out, from the consideration of what we already know, something else we do not know.'"

"That's beautiful," Berkowitz said appreciatively.

"It is, isn't it? But the problem is, if reason is a human invention, in other words a process that takes place according to rules of logic that we make up, using language that we also make up, how can it tell us anything about the world outside? It can only talk about things within the context of its own closed loop of words and definitions and rules."

"Just a minute," Berkowitz said. "I'm not sure I . . ."

"Think of it this way," I said. "Reasoning consists of combining fact A with fact B, according to a set of rules, in order to produce the previously unknown fact C. The question is, How does this happen? What makes C emerge? Does the process of combining A and B 'cause' C to emerge, like a chemical reaction? In other words, is C something entirely new? Or are A plus B on the one hand, and C on the other, simply translations of one another at different levels of knowledge?

"Let's say that C is genuinely new knowledge, just for the sake of argument. Then it would be possible to combine C with D, to produce E, at another, higher level. That way, reason should be able to build its accumulated base of knowledge all the way to the stars, right?"

"Right," said Berkowitz.

"But there's a problem."

"I thought there might be."

"The problem is this. At its most fundamental, foundational levels, this edifice of reason is floating in thin air. Its basic A and B assumptions have no contact with reality. They're just ideas we've made up."

"Well? How *does* reason, or science, connect with the real world?" Berkowitz wanted to know. Sister Celeste raised a finger to indicate she had an answer.

"The answer, I think, is that it does not! What science does is build models. Its basic construction material is mathematics. The model is like a map. If it is successful, it will help us to negotiate our way around in nature. If it is not, we will be, as it were, falling off cliffs and bumping into mountains. We must always remember that what we are seeing in our experimental results is not the territory but the map. Or we might say that the model's connection to the reality it describes is analogous to the link between words and the objects they denote. The word *chair*, I think we would agree, is not the object."

"I still don't see . . ." Berkowitz began.

"Ah yes, you want to know how all this beautiful logic applies in the real world. For me, what it means is that if our scientific understanding of the nature of reality is merely a model that we ourselves construct, then it is conceivable—probable, even—that other models could be constructed that would serve us as well. I can imagine various models of the world that would be 'objective' in the sense that they would allow us to function well in our environment but might be particularly useful for various purposes. One, such as the current model, might keep us from blowing ourselves up when combining certain chemicals, while another might keep us from murdering each other over imagined slights to our honor, or a third might keep us from poisoning our environment. A fourth might conceivably do all three.

"Allow me to return to our topic of the other night, for a moment. What makes science so successful, as we discussed, is its single-mindedness. In other words, the model it has of the world is very small and very well defined. It deliberately excludes consideration of most of the things that make life worth living, and confines itself only to what can be counted and measured. This almost guarantees that it will be successful, because it refuses even to consider the realm of information that might lead to a questioning of that success. It will have no truck with intangibles.

"Science is successful because it refuses to deal with the difficult questions of the meaning and purpose of things, the meaning of our lives, for instance, and the purposes of nature. It occupies itself full-time tinkering with its rather desiccated little model. Science knows that Newton one day had a bright idea and called it gravity, but it knows nothing about how gravity informed Newton of its existence, or what gravity *is*, or *why* it is."

"It seems to me, Sister," I said, "that your idea of different models all having validity is given credence by the scientific model itself. Because that model from time to time is scrapped in favor of a new one. We had the Ptolemaic-Aristotelian model throughout the Middle Ages until Galileo and Copernicus, and then we had the Newtonian model from the seventeenth century right up until the beginning of the twentieth century, when that was replaced by quantum mechanics and Einstein's relativity. And there's no good reason to think that that will be the last word. The world 'worked' for people in all these models, and yet they are quite different from one another."

Berkowitz interjected impatiently, "Yeah, but that's the great thing about science. That's the march of scientific progress you're talking about. One idea builds on another. The great scientists stand on the shoulders of those who went before, or whatever that saying is. Quantum mechanics didn't replace Newtonian physics, it just added on to it."

I responded that I didn't think he would find many scientists to support that view. "Scientists would say, I think, that Newtonian physics is simply wrong. It works at the human scale of things for most purposes, and in that sense it is a good model. But it does not work at either the small scale of subatomic particles or the large scale of interstellar space. It just happens to provide some pretty good approximations in the bit of the universe we occupy. Even at the scale of our own solar system it does not really work, because there are planetary motions that it cannot account for, that can only be accounted for within the quantum mechanical, relativistic model. Newton's scheme of things was like one of those illustrated tourist maps you find on paper placemats in highway restaurants. They'll give you a rough idea of where A is in relation to B, but they're useless once you get off the beaten track. They are not what you would call a real map."

Berkowitz washed down the last of his chocolate-filled croissant with his coffee. "So what you two are telling me is that any number of models of reality will do, and that none of them has any discoverable link to things as they really are, beyond what works, which can only be determined by trial and error?"

"I would not put it quite in that way," the little nun said, wagging an index finger. "I believe that some models *do* have a very real link with reality 'out there'; and these are the religious models. I would go so far as to say that the purpose of religion is precisely to provide that otherwise missing link between our perceptions and the underlying reality. The link is to be found in revealed wisdom, revelation. We Christians have a record of it in the Holy Scriptures. A Confucian or a Moslem might say the same."

"But you can't prove . . ." Berkowitz began, leaning forward over the table, and then he deflated visibly and slumped against his chair back.

"Yes, and neither can science, Mr. Berkowitz, as I think you agreed yesterday. And so, as Mr. William James suggests—your fellow Pragmatist, Mr. Berkowitz, and not a churchgoing man—it makes good sense to suspend disbelief about religious knowledge and try to meet it halfway. It makes sense for all of us, including, and perhaps especially, scientists and the scientific-minded. They need not throw out their existing model, but they may well discover ways to extend it usefully and to modify its rules of operation so that it better reflects the world as most ordinary people know it. It seems to me that faith and scientific conjecture are functionally the same with respect to learning: they can both provide our reason with—what shall we call them?—*presuppositions* from which it can arrive at bits of knowledge. There is a lovely poem by Robert Frost . . . let me see if I can remember it."

She closed her eyes for a moment and then began to recite:

Others taunt me with having knelt at well-curbs
Always wrong to the light, so never seeing
Deeper down in the well than where the water
Gives me back a shining surface picture
Me myself in the summer heaven, godlike,
Looking out of a wreath of fern and cloud-puffs.
Once, when trying with chin against a well-curb,
I discerned, as I thought, beyond the picture,
Through the picture, a something white, uncertain,
Something more of the depths—and then I lost it.

The little nun laughed.

"Dear me, I sound like an evangelist, don't I? I am afraid, though, that I really must be on my way. My parents are getting somewhat frail, and I promised I would accompany them to the market in the Rialto."

I stood and held her chair. Berkowitz remained slumped in his place, hands tucked into his armpits, legs extended beneath the table, crossed at the ankles.

"You've been very kind, Sister," I said. "We thank you."

"Yes, yes, thank you," said Berkowitz, forcing a smile.

We watched her cross the lobby to the entrance door.

"You're not your usual gracious self," I said.

"She has all the goddamn arguments," he grumbled.

"Why not go with it, then?"

He stood up abruptly, angrily, his napkin falling to the floor. "Can we leave?"

I had an idea what was coming. I paid the bill with cash, and we left the hotel. We walked in silence for several minutes, along the seashore where the roofs of hundreds of brightly painted bathing cabins peeked over the embankment, lined up like soldiers. As we reached the Grand Excelsior, he suddenly began berating me.

"You give me all the crummy lines, all the feeble arguments! I'm humiliated every time I open my mouth! What's more, you make me out to be a total boor—"

"Listen," I said, raising my voice, talking over him, "you can't lay all of that on me. You have to take some responsibility for your own behavior. You know perfectly well I don't just make that stuff up out of thin air—"

"I tell you, I'm fed up with being the simpleton of the piece and I'm fed up with being made to look like a total jerk. You and your precious nun prattle on sanctimoniously about the evils of science and the blessings of spiritual values, nobody challenging you. Well, let me tell you, I know a thing or two about history, and the Age of Reason and the Reformation weren't welcomed as liberating events for nothing! People were fed up with being slaves to intellectual authoritarianism—terrorism! People's minds were in cages, and they lived in barnyard conditions. Is that what you want to go back to?"

"Finished?" I asked. We stood facing each other on the sidewalk.

"As if I have a choice."

"Fine," I said, taking his words as a response in the affirmative. "Number one: I don't give you 'crummy' arguments. I give you the very best arguments available to your side. If you want winning arguments, you're out of luck. You've appeared in the wrong century. No serious thinker has bought into the empiricist, mechanistic point of view for at least a hundred years. It's a Potemkin village. Unfortunately, that news has yet to seep through to the general public. The scientistic view is almost never examined seriously and it's still considered 'common sense.' If you dare to challenge it, its supporters declare a jihad. But when it comes right down to it, the idea of the world as a kind of giant pinball machine that accidentally produces everything that we call nature doesn't hold water scientifically, philosophically, or even in terms of every day experience.

"Number two: you talk about caged minds—a nice turn of phrase, by the way—but I'm telling you the human mind was caged when it disengaged itself from a collective vision and became the mind of the sovereign individual in the Age of Reason. Or maybe it happened before that, with the invention of the alphabet and the rise of literacy, as McLuhan says. *That* is a caged mind—a mind that tries to wall itself off from those around it. That is a mind open to exploitation and manipulation because it has only its own resources and defenses to fall back on. The *liberated* mind is the one that has access to and can navigate the collective intelligence of the human community. But you don't get that access without buying into the network and to do that you need to understand and accept that it exists, that the human race is more than an aggregation of singularities, of self-serving automatons. The caged mind, my friend, is the mind of the logician and the rationalist, where everything is mutually exclusive, black or white, right or wrong, fact or fiction, one or zero. That is a mind that is trapped in arithmetic, unplugged, disconnected, onanistic . . .

"As for the arguments, you're just going to have to be more flexible, more amenable to persuasion. More professional. You can't win, any more than Simplicio could win against Salviati in Galileo's *Dialogues*. So you might as well fight the good fight and lose gracefully, as Simplicio did. You'll at least have the comfort of knowing you lost to the superior position."

Berkowitz turned and strode on, shoulders hunched, fists jammed in pockets, eyebrows at maximum declination. But as we neared the dock, his posture straightened and the tension seemed to melt from his face. He appeared to positively enjoy the vaporetto ride back.

SEVENTEEN

At Santa Maria Maggiore

Galileo's Conservatism • Kepler's Heterodoxy

A Remarkable Fresco

Shortly after my return from Venice, I took a long walk under lowering skies from my hotel to the great church of Santa Maria Maggiore on the Esquilino Hill. The church was founded by Pope Sixtus III (432–40) and contains some of the finest fifth-century mosaics to be seen anywhere. Its coffered ceiling is gilded with the first gold to be shipped to Europe from Peru, a donation from Ferdinand and Isabella of Spain more than a thousand years later. I had arranged to meet Berkowitz there, not to look at the mosaics or the gold but at one of the most remarkable ceiling frescoes in Rome.

I was by no means certain he would keep our appointment and was much relieved when I spotted him from a good distance away, near the top of the steps that led to the front entrance to the church, apparently deep in conversation with three priests dressed in black slacks and shirts. All four were gesticulating extravagantly, and the movements of their arms gave the impression of some undersea creature capturing its dinner. Before I could reach them, the group broke up abruptly, the priests moving off down the stairs talking among themselves, leaving Berkowitz, arms akimbo, staring after them. When he finally noticed me approaching he seemed to brighten and waved a hello.

I climbed the last few steps to where he was standing. "What was that all about? Looking for directions to a recruiting office, were you?"

"Fat chance!" he snorted. "They were speaking English, and so I introduced myself. You know, trying to spread international good will and all that. I mentioned the fresco inside, and they immediately started dumping on Galileo. He's practically the devil for causing the Church such grief. I reminded them that PJP2 has rehabilitated Galileo . . ."

"PJP2? . . ."

"Pope John Paul the Second. In the encyclical 'Faith and Reason' he says the Church was wrong about Galileo. They say Galileo is mentioned only in a footnote. Well, I only know what I see in the newspapers, and that's what I saw. Then they say that Galileo wasn't much of a scientist anyway, at which point I have to laugh. Kepler, they say, had a better grasp of things. Kepler! Who the hell is Kepler!? I mean, I know who he is, but he's hardly in the same league as Galileo, or he'd be a household name, right? All of this happens in about two minutes. The tempers these people have!"

"Well, it is true that Kepler solved the problem of planetary orbits, and Galileo didn't," I offered.

"What are you talking about? Copernicus figured that out."

"Copernicus had the planets orbiting the Sun, all right, but he had a Rube Goldberg arrangement almost as complicated as Ptolemy's, complete with epicycles and equants and all that. The reason is that he still believed that orbits had to be perfectly circular. Galileo did, too, and he subscribed to the Copernican system with all its complexities and circular orbits till the day he died. It was Kepler who saw that the orbits had to be elliptical."

"Well, obviously, Galileo didn't hear about that."

"Oh, but he did, and straight from the horse's mouth. Kepler sent him all his publications, and asked him for feedback. He informed Galileo about elliptical orbits more than once."

"And what did Galileo say?"

"Nothing. He never replied to Kepler's requests for comment. It's one of the big mysteries of the history of science."

"What? That he didn't reply? Postal systems were pretty dodgy back then . . ."

A confident Galileo in middle age, c. 1610.

The Council of Trent as depicted in a sixteenth-century allegorical fresco in the basilica of Santa Maria in Trastavere, Rome.

Nicolaus Copernicus (1473–1543) introduced a revolutionary system of planetary motion.

Johannes Kepler (1571–1630), astrologer, astronomer, mathematician.

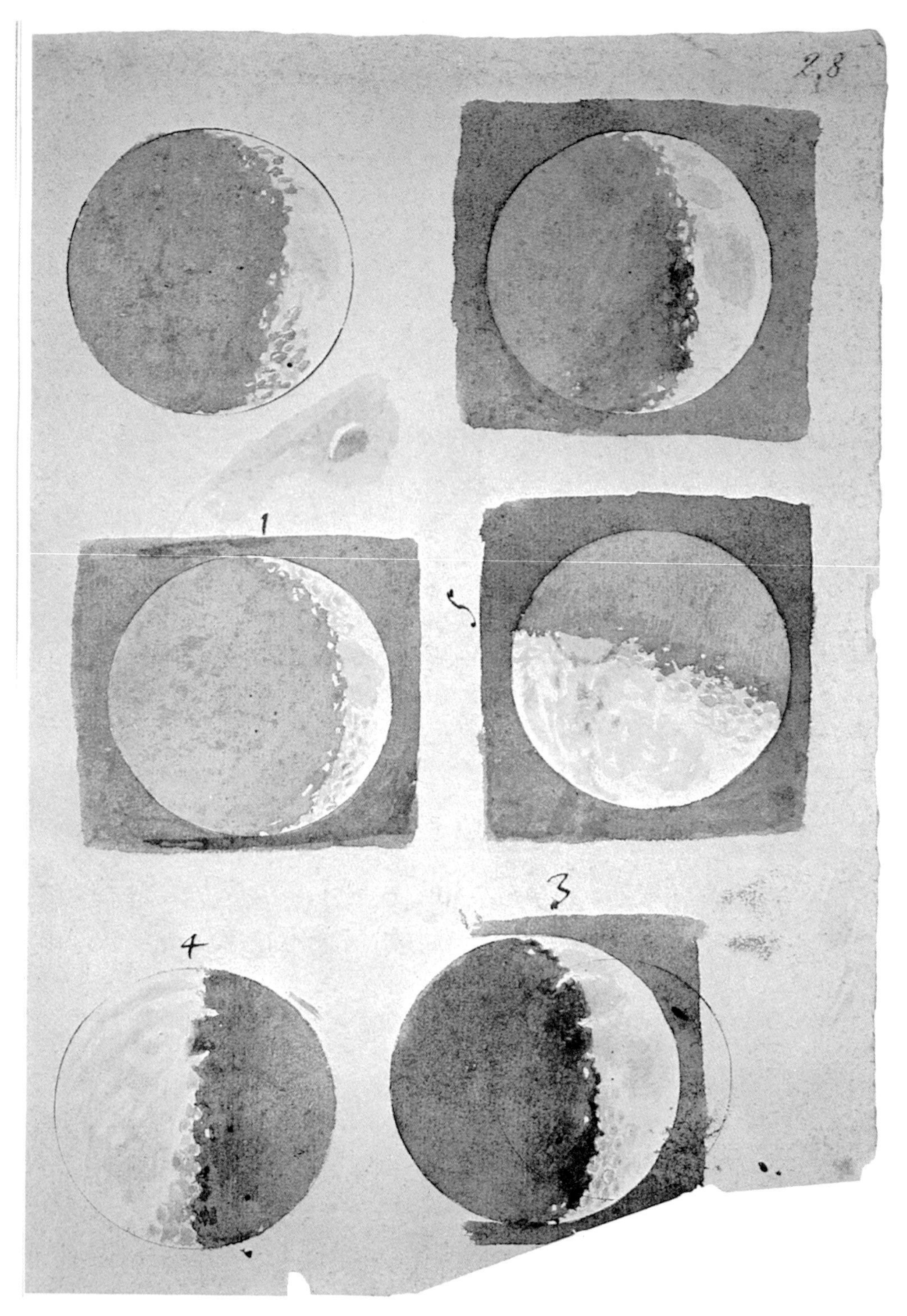

Galileo's sketches of the Moon as he saw it through his telescope.

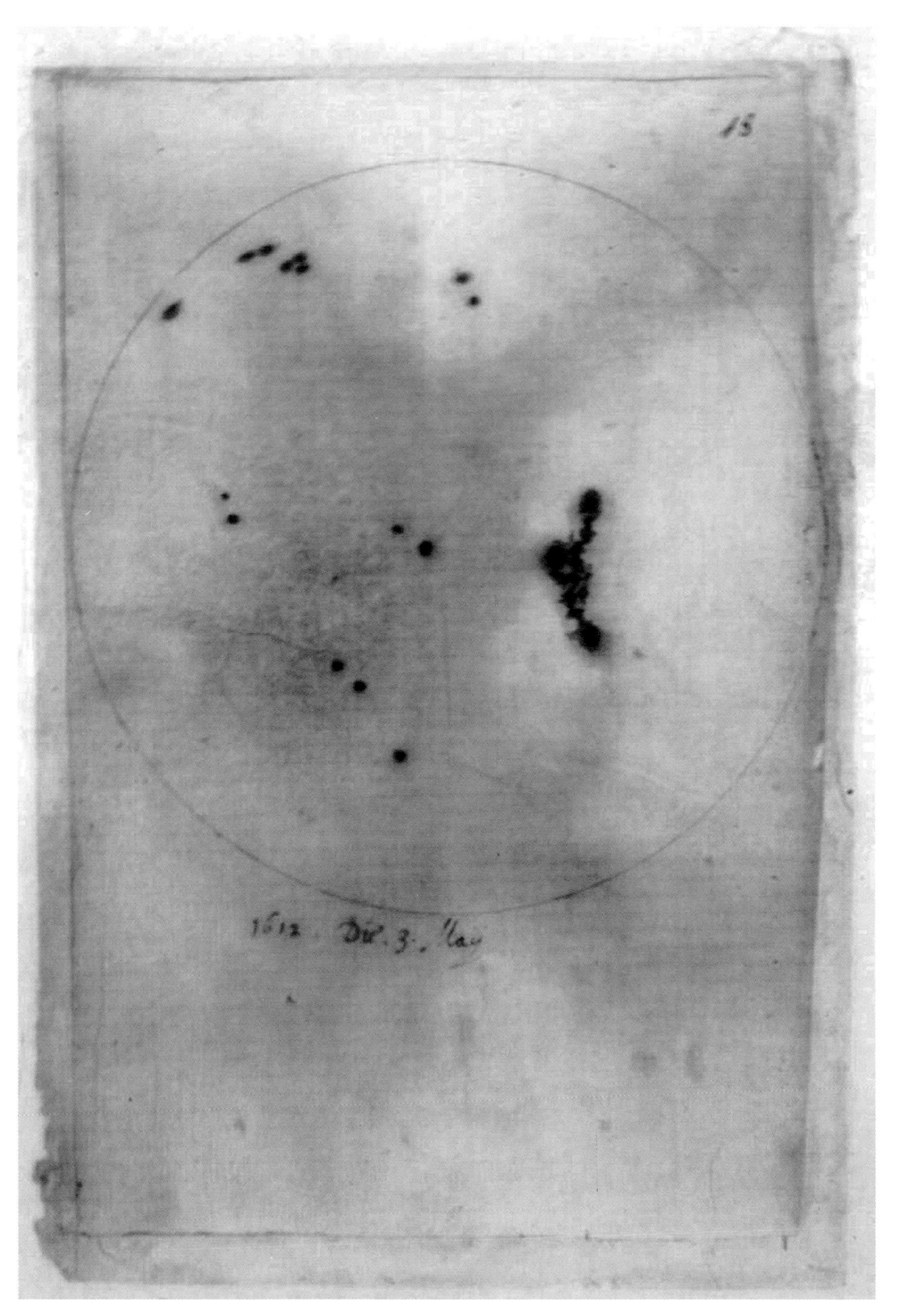

A sketch of sunspots by Galileo.

Ludovico Cigoli's ceiling fresco in the Pauline chapel of the church of Santa Maria Maggiore.
The moon beneath Mary's feet is Galileo's moon.

Camilo Borghese, Pope Paul V, whose reign saw the dawn of the Scientific Revolution.

Matteo Barberini, Pope Urban VIII, was pontiff during Galileo's trial by the Inquisition.

Galileo confronting the Inquisition's commissary general, in the nineteenth-century representation by Cristiano Banti.

Galileo's tomb in the church of Santa Croce, Florence.

"The mystery," I said, "is that he didn't take up the idea of elliptical orbits. It solved all the problems with Copernicus and got rid of all the complications at a stroke. Why wouldn't he have adopted the idea?"

We had reached the top of the stairs and were now standing outside the great entrance doors of the church.

"Well?" Berkowitz was favoring me with his most penetrating gaze.

"Well, some historians put it down to Galileo's ego, which would have filled the Pantheon. They say he just couldn't accept that somebody else's ideas—some competing astronomer's ideas—were better than his own. But I don't think that rings true. He was too much a scientist for that. Others say when he wrote his *Dialogue on the Two Chief World Systems* and left out ellipses he was trying to make it as widely accessible as possible, and simply didn't want to include the added complication of elliptical orbits. But then, he never mentioned elliptical orbits in any of his private correspondence either, so that pretty well rules out that argument. It's a puzzle. The best explanation I've seen is one Erwin Panofsky suggested."

"The art historian?"

"None other. He wrote an obscure little pamphlet called 'Galileo as a Critic of the Arts' back in about 1954. In it he argued that it was Galileo's aesthetic attachment to the circle that prevented him from going with the ellipse."

"So what he's saying is that Galileo thought that the best scientific solutions are always the most elegant, the ones that display the most beautiful symmetry and so on."

"Right, but there's more to the argument. Shall we go in, or—"

"No, no," Berkowitz said, "I hate talking in churches. Makes me feel like a heathen. Which I am."

"Well, then, Panofsky is fairly convincing as far as he goes, but I think that there's a more hard-core philosophical reason for Galileo's attachment to the circle as well. And there's a big irony in this one. Galileo studies mechanics because he understands that the cornerstone of physics is motion, and he thinks Aristotle's ideas about motion are rubbish, which they mostly were. One of the biggest problems he sees with Aristotle is that his system allows for straight-line motion. Air and water, for example, can move straight up, indefinitely. I'm talking about what we would call inertial motion, and what Galileo—and Aristotle—referred to as natural motion. In other words, the way bodies move if they're left to themselves. Galileo believed very strongly that all natural motion is circular, and he says so right at the beginning of his *Dialogue.*"

"You're pulling my leg," Berkowitz said. "All natural motion is circular . . ."

"Yes, and he had a pretty good reason why it had to be. He said that if a thing naturally moved in a straight line it would carry on doing so forever, right into infinity. But that would mean the universe must be infinite.* And Galileo, like virtually all his contemporaries, not to mention just about everybody else for the previous two thousand years or so, was quite certain that the universe could not be infinite. The word for the universe is *cosmos*, which means order. An infinite universe could not be ordered, because there would be no up or down, no center, no points of reference of any kind. It would be the opposite of order, which is chaos. The circle maintained order, while the straight line allowed things to move ever farther apart into disorder.

"You know what the moral of the story is?" I asked Berkowitz.

"Keep your voice down," he whispered. "What?"

"The moral of the story is that Kepler pursued his investigations of the solar system from a position that today's science considers completely off the wall—Pythagorean mysticism, astrology, numerology, Christianity. And yet he came up with the right answers when Galileo, the father of modern science, could not.

"Which demonstrates that there is more than one valid way to do useful science. You don't have to be a mathematical realist—a positivist like Galileo and the rest of his tribe—to do good science. You don't have to buy into the scientistic worldview. In fact,

---

*This is the way Galileo puts it, in the words of Salviati in the *Dialogue*: ". . . [T]he straight motion being by nature infinite and indeterminate, it is impossible that any movable body can have a natural principle of moving in a straight line, namely, toward the place whither it is impossible to arrive, there being no predetermined limit; and Nature, as Aristotle himself well says, never attempts to do that which cannot be done or to move whither it is impossible to arrive." (Kepler, Epitome IV, 3.1, quoted in Panofsky, "Galileo as a Critic of the Arts" [St. Martinus Nijhoff (1954), 30]).

sometimes it can help if you don't.

"Furthermore," I said, "that's the reason why Galileo is a hero and nobody's ever heard of Kepler. Kepler is a scientific heretic. His views are heterodox. So he's been suppressed, marginalized. If he were alive today he'd be drummed out of the profession, cut off from grants, denied a teaching position, out of a job . . ."

"For God's sake, get a grip," Berkowitz hissed. "Keep your voice down!"

I could tell from the way he avoided eye contact that I'd scored a point.

We walked up the nave between ranks of polished marble columns on Cosmati pavement, a fine mosaic inlay of marble, glass paste, and gold leaf. I recalled from somewhere that the Cosmati school was so called because many of its members had the given name Cosmos, or Cosimo in Italian. I wondered whether anyone still named their sons for the cosmos. All along the entablature supported by the columns were fifth-century mosaics depicting stories from the Old Testament. It was in that century that Rome was repeatedly pillaged by the Vandals and Visigoths and Ostrogoths. How could such beautiful work have been done under such conditions, I wondered? Like so many other buildings in this marvelous city, it occurred to me, this church had a biography that could tell much of the history of Western civilization.

The painting I sought was a fresco on the ceiling of the Pauline chapel, also called the Borghese chapel. It was commissioned by the Borghese Pope Paul V, a man familiar, of course, to Galileo. It will be recalled that it was this "anti-intellectual" pope's advisory commission that issued the report declaring Copernicanism scientifically untenable and theologically heretical. Following on that, Galileo was warned not to teach or defend the hypothesis, and from there unfolded the events that led to his eventual summons and trial by the Inquisition.

How ironic, then, that this painting should be found in the chapel Paul V commissioned for himself in one of the greatest churches in Rome! For if you look up to the ceiling where Ludovico Cigoli has painted a scene of Mary being transported to heaven, you will see her foot planted firmly on the Moon. And the Moon is not the perfect, luminous sphere of Aristotle, but the cratered, mountainous

Moon revealed by Galileo's telescope, complete with its ragged terminator where the Earth's shadow cuts across the rugged lunar surface.

"Notice anything funny about the Moon?" I asked Berkowitz, as we stood looking upward into the chapel dome.

"It looks like the plaster is blistering," he said. After a moment he added, "And it looks like the top of some kind of crumbling stone pedestal. It's not spherical. What makes you think it's the Moon?"

"Come stand over here. Look closer," I said, irritated by his mulishness. I pulled him farther into the chapel, where the foreshortening effect was absent, and the painting appeared more as it might on a flat surface.

Berkowitz stood looking straight up for a long time, his jaw pulled open by the stretched skin of his neck. Finally he looked down again.

"All right, so it's the actual Moon, warts and all. What's your point? I don't think it was unusual to depict Mary standing on the Moon, was it?"

"No, but remember that it was this particular pope who started all the trouble for Galileo and that the painter, Cigoli, was one of Galileo's best buddies. This was his last major commission. Mary's foot is on the Moon to show she's queen of the universe. But whose universe? That's Galileo's Moon! It's not Aristotle's perfect Moon up there, and it's not Paul's either. That Moon, don't you see, is a finger in the eye of the pope and all his fellow Aristotelians!"

There was a second or two of silence, and then Berkowitz let out one of his trademark stentorian cackles and immediately clapped a hand over his mouth. The echo seemed to reverberate endlessly.

"That's beautiful," he whispered when he'd recovered his composure.

"I thought you'd enjoy it," I said.*

---

*Looking at the fresco in its current damaged condition may lead to doubts about whether it is indeed the Moon beneath Mary's foot. This is especially true if one does not stand in the correct position to minimize the foreshortening—which would not have been so troublesome if Cigoli had mastered perspective as thoroughly as contemporaries like Andrea Pozzo. However, the fact that the painter depicted the Moon in the fresco is confirmed by at least two contemporary letters, which are quoted in Panofsky's pamphlet. A further convincing documentary reference is found in Roberto Contini's book *Il Cigoli* (Edizione dei Soncino [1991]).

EIGHTEEN

A Scientific Testament • Models of the World
The Paradox of Empiricism

Following the February 1616 decree of the Congregation of the Index and the private warning from Cardinal Bellarmine, Galileo was as good as his word, and for several years refrained from teaching or publicly advocating the Copernican hypothesis. He focused instead on quietly building the scientific case for the Earth's movement.

Early in 1618 he was busily at work revising a paper on tides he had shown to friends in Rome. His conviction was solidifying that in tidal action, which he believed was caused by the motion of the Earth, he had found the conclusive proof he needed to force Church authorities to reconsider their stand on Copernicus. He wrote a prudent preface describing the theory as merely an ingenious speculation and sent the manuscript off to Archduke Leopold of Austria, perhaps hoping it would be published in that country.*

*Younger brother of Emperor Ferdinand II of the Holy Roman Empire. In the context of the complex and highly competitive politics of patronage, Galileo had other reasons for cultivating Leopold. He was the brother of the wife of Galileo's patron, the duke of Tuscany, and his favor would help to ensure Galileo's continuing esteem in the Tuscan court. Galileo may have failed to consider that Urban VIII was in a continuing state of eggshell diplomacy with the Habsburg courts of Spain and Austria and their allies (which included Tuscany) and regarded them as the Vatican's major European threat. Galileo sent the archduke a number of expensive gifts, including several telescopes and a copy of his *Sunspot Letters*.

And then in the same year events occurred that led him to write and publish a work that is at the same time a classic of sarcastic rhetoric and critical invective and one of the great works of scientific literature, *Il Saggiatore (The Assayer)*. Three comets appeared in the autumn sky over a period of a few weeks, the last and brightest of them remaining clearly visible well into winter. Galileo was suffering one of his periodic bouts of illness and was bedridden much of the time, so spent little time observing them. In any event, he did not believe comets to be true celestial objects, but merely atmospheric vapors ejected from Earth to vanish at great distances.

Tycho Brahe had more than thirty years earlier established by parallax measurements that the comet of 1577 was located far beyond the Moon, and he considered it to be a heavenly object. The Jesuits at the Collegio Romano had accepted this view, as well as Tycho's conclusion that comets had orbits like planets. But the orbits were discovered to be so markedly elliptical that when the 1618 comets appeared, a teacher of mathematics, Father Horatio Grassi, delivered a lecture at the Collegio arguing that they were the best possible evidence against Copernicanism (which, as we know, postulated circular orbits). This was subsequently published as a pamphlet, "Discourse on Comets." It should be remembered that while the Jesuit astronomers had, like Galileo, abandoned the ancient Ptolemaic system in the face of telescopic discoveries, they had rejected Copernicanism for both scientific and scriptural reasons. Instead, they favored the model proposed by Tycho Brahe. The Tychonic system preserved an unmoving Earth at the center of the planetary system, but the other planets revolved around the Sun, which in turn orbited the Earth. It was mathematically simpler than the Ptolemaic system and provided more accurate predictions.

Galileo's friends, especially those in the Lincean Academy, egged him on to respond to Grassi, and it was that response that would eventually be published by the Academy as *The Assayer*. Galileo was of course prevented by Bellarmine's injunction from making direct references to the Copernican position: instead, he attacked what in the circumstances can only be called a straw man. It was his perennial hobgoblin, the validity of relying on authority (in this case, Aristotle) as a sound basis for knowledge. As an alternative he set out in detail the approach he had been working with since his earliest mechanical experiments at the universities of Pisa and Padua. For that reason it remains one of the foundational documents of modern science.

Until Galileo, scientific reasoning had been done deductively: Galileo proposed an inductive method. I can do no better in explaining the difference than to repeat a short passage from Sister Celeste's essay on Aristotle, the champion of the deductive method:

> [Aristotle's science] was not a science that encouraged discovery in the sense in which we understand it, but rather a deep study of what was already known. The undeveloped was to be explained in terms of the fully developed, and not, as in modern science, the other way around. The plant, the achieved form, is prior to the seed, as the man is prior to the child. Aristotle's science reasoned deductively, from the top down, from the complete to the incomplete, as opposed to modern science's inductive or bottom-up approach, in which phenomena are broken down into their most basic elements, each of which is studied separately.

Galileo's innovation went beyond a preference for induction. The basic data from which he built his scientific inferences, he insisted, had to be gained by actual experience of the real world. By this he meant either direct observation or, where possible, experiment. In astronomy, experiment was clearly out of the question and so observation had to suffice, but in the realm of mechanics, experiment was the preferred route. His system of knowledge, as a philosopher might say, was built on a foundation of empirical evidence. With this he combined mathematical interpretation to produce in all its basic elements what we know today as the scientific method.

While others before him had used experimental observations to disprove Aristotle's assumptions, Galileo's experiments went beyond this to provide him with the data he needed to prescribe the correct mathematical relationships at work, so that any new hypothesis could be tested on paper. Galileo lifted geometry out of its confining and too-concrete environment of lengths, areas, and volumes, made it abstract, and began applying it to other areas of natural phenomena such as time, motion, and mass so that he could examine the relationships among them. He began making mathematical models of the world. In doing so he deeded science both a powerful vision and a dangerous blindness.

To make his models, it was necessary for him to ignore anything that could not be reduced to number and concentrate only on those things that could. Only qualities that were measurable were worthy of observation for his purposes. Nonmeasurable qualities, in other words, had no place in Galileo's science—or, indeed, in the science of his successors down to the present day.

Measurable qualities he called "primary qualities," and those that were not measurable he called "secondary." In the first category were such qualities as mass, motion, and volume. These were regarded as objective, real properties of matter. In the second category were such qualities as taste, smell, and color, which were treated as subjective creations of the sense organs, possessing no reality in the external world. Or, as he put it in *The Assayer*, "Many sensations which are supposed to be qualities residing in external objects have no real existence save in us, and outside ourselves are mere names."[1]

Even further beyond the pale were moral qualities such as good and bad or just and unjust. For Galileo, these fell into a similar category, and were "in no way superior, to the 'sympathy,' 'antipathy,' 'occult properties,' 'influences,' and other terms employed by some philosophers as a cloak for the correct reply [to a scientific puzzle], which would be: 'I do not know.'"[2] In Aristotelian and Christian physical science, these moral qualities were important elements simply because science was an integrated part of the overall philosophical systems. The teleology, or goals, of Christianity were part and parcel of Christian concepts of science, so it was assumed that physical processes were ultimately tied up with the best interests of humanity and therefore contained moral qualities. This ended with Galileo, and henceforth all nonprimary qualities came to be regarded as unreal.

What this meant in practice was that science would be swept clean of theoretical knowledge that did not conform to the mathematical models developed out of experiments like Galileo's. In the past, conflicting, even contradictory, theories and hypotheses had existed alongside one another for centuries. Now they could be tested and rejected with relative ease. It was now possible, as Galileo pointed out, to demonstrate "what perhaps has never been observed" from formulas derived from phenomena that were known. The theoretical demonstration, he said, provided an explanation for those new phenomena, and the existence of phenomena themselves provided verification of the explanation.

But there is a fatal problem here. The process of verification is a circular one, unsound in logic.* The insights of reason are like words in a dictionary, each of which depends on some other entry in the same dictionary for its definition. Humans live mostly inside the covers of that dictionary. Ultimately, every insight based on reason depends for its premises on its own conclusions. As creatures of God and nature, human beings cannot, as a matter of routine, see outside that closed, recursive system. Their reason, confined between the covers of the dictionary, cannot make contact with the reality behind the definitions. Human reason can show us only *imagery* of a more profound reality: it can show us the words, but not the things they represent. We can know with a high degree of certainty that a more profound reality is out there, because why else would the dictionary exist? But the worlds we build using that imagery—the worlds that science describes—are not real in any ultimate sense but merely metaphors for that deeper underlying reality.

To restate the problem in slightly simpler terms: Galileo argued that the mathematical hypothesis explains the phenomenon, and the phenomenon's existence verifies the hypothesis. But this can be true if and only if an unspoken assumption is made. And that is—as Galileo would join Pythagoras in insisting—that the

---

*Galileo was aware of this problem, though perhaps not at the time of writing *The Assayer*. In the fourth day of the *Dialogue* published nine years later he raises it through the words of Simplicio during a discussion of the causes of the tides: "Under the assumption of the two terrestrial movements, you give reasons for the ebb and flow, and then, vice versa, reasoning circularly, you draw from the ebbing and flowing the sign and confirmation of those same two movements." Salviati, or Galileo, has no response to the charge, which is essentially unanswerable.

fundamental relationships in nature are numerical. However, there is no reason in logic to assume that because a mathematical model predicts a certain phenomenon, and that phenomenon is subsequently found to exist, that the mathematical model is a sufficient or necessary or even adequate explanation of the phenomenon. The phenomenon could have been caused by some agency completely outside the limits of the model. Alternatively, the model's ability to "predict" might be pure coincidence, or it could be a result of the model's approximating, either by design or accident, the workings of nature at a deeper level.

As early as the eighteenth century, however, philosophers like David Hume and Immanuel Kant were questioning the foundations of the scientific approach introduced by Galileo. The debates among philosophers and scientists continued through the nineteenth century and into the twentieth. And in the late twentieth century scientists themselves were beginning to have doubts. By that time Einstein's relativity had confronted and swept away Newton's physics. Newton's "laws" continued to work—they still provided useful predictions in everything from designing spacecraft to riding bicycles. But it was now seen that they were only mathematical approximations, and the world is actually a lot more complicated and a lot less predictable than Newton had believed it to be. Newtonian physics works for most purposes here on Earth, at the scale on which we humans normally function. But beyond Earthin the broad reaches of the universe, and in the micro-universe contained within atoms, parallel lines can and do meet; mass and energy are interchangeable; reactions can happen before the action that "causes" them; there are many more than four dimensions; fundamental particles seem to have no real existence beyond the mathematical formulas that describe them; matter may be nothing more than consciousness, or information, or a combination of the two; things travel faster than the speed of light; time is reversible . . . and so on.

But in the first flush of enthusiasm, mathematical models appeared to Galileo and his supporters and successors to be the long-sought key to unlocking the universe. "Philosophy is written in this grand book the universe, which stands continually open to our gaze," Galileo said in *The Assayer*, "but the book cannot be understood unless one first learns to comprehend the language and to read the alphabet in which it is composed. It is written in the language of mathematics, and its characters are triangles, circles, and other geometric figures, without which it is humanly impossible to understand a single word of it; without these, one wanders about in a dark labyrinth."[3]

He saw himself as one of a small number of intellectually gifted modern philosophers using a superior method for the acquisition of knowledge. His understanding of nature, he believed, was illuminated by the light of mathematics, so he had no need to rely on authority. It was possible for humankind, using the scientific method, to have knowledge of specific areas of interest every bit as precise as God's (though he conceded the scientist could not match God for *breadth* of knowledge). For what could be more absolute and definitive than mathematical demonstration of a natural phenomenon?

Addressing the Jesuit mathematician Horatio Grassi (who wrote under the pseudonym Lothario Sarsi) in *The Assayer*, he makes his claim to intellectual ascendancy in a self-important way that seemed calculated to earn the animosity of his former allies at the Collegio Romano:

> Sarsi [Grassi] perhaps believes that all hosts of good philosophers may be enclosed within walls of some sort. I believe, Sarsi, that they fly, and that they fly alone like eagles, and not like starlings. It is true that because eagles are scarce they are little seen and less heard, whereas birds that fly in flocks fill the sky with shrieks and cries wherever they settle, and befoul the earth beneath them. . . . If reasoning were like hauling I should agree that several horses can haul more worth than one, just as several horses can haul more sacks of grain than one can. But reasoning is like racing and not like hauling, and a single Arabian steed can outrun a hundred plowhorses. So when Sarsi brings in his multitude of authors it appears to me that instead of strengthening his conclusion he merely ennobles our case by showing that we have outreasoned many men of great reputation.[4]

The Assayer was just going to the printer when Galileo's old friend Maffeo Barberini was elected Pope Urban VIII. The Linceans hurriedly redesigned the title page to include a dedication to the new pontiff, who was by all accounts amused by the book, which he had read to him at table.

NINETEEN

Pope Urban VIII • Intimations of Change
Galileo's Dialogue *• The Suppression*
A Summons from the Inquisition

In the spring of 1624, Galileo set out for Rome to join in the papal election celebrations. He knew by then that the gift of his new book, *The Assayer*, had been gratefully accepted and enjoyed, and he could look forward to a cordial welcome. Soon after he had installed himself and his retinue at the Villa Medici, he was granted the first of six long, weekly audiences with the pope.

While there is no record of what the two discussed, we can surmise that the scientist pressed his argument that it was dangerous for the Church to maintain doctrines and scriptural interpretations that flew in the face of mounting physical evidence of Copernicanism. The pope, in turn, can be expected to have cautioned Galileo with his own skepticism about scientific knowledge, a position he based on the omnipotence of God, and one he had explained in detail to Galileo as early as 1616, at the time of the temporary Indexing of Copernicus. There was no such thing, he believed, as scientific necessity or immutable laws of nature, because their existence would restrict the power of the Deity. Where we see such laws, he believed, we are merely projecting human rationality onto a fundamentally mysterious process, in order to make sense of it. Thus, it was wrong to say that the Copernican hypothesis, or any other scientific theory, represented truth in any deep sense. And so, as Urban would concede to Galileo, there was nothing wrong with treating the Earth's motion as a hypothesis and studying the consequences for the science of astronomical observation and prediction. Urban did not think Copernicanism a heresy but rather a dangerous doctrine that ought to be treated with circumspection. As he doubtless reminded Galileo, he had fought to prevent Copernicus from being placed on the banned list of the Index in 1616. And having said that, he is unlikely to have omitted the fact that without his personal intervention with Paul V at that time, Galileo himself would likely have been dragged into the controversy and perhaps even disciplined in some way.

All things considered, the new pope seemed a breath of fresh air to Galileo, who left Rome laden with gifts from Urban, including two medals, a painting, and a Church pension for his son, Vincenzo. Later, he would be heartened to hear that the pope had appointed two of his former pupils to important posts: the Dominican Niccolò Riccardi was named master of the Apostolic Palace, whose responsibility it was to authorize publication of books, and Benedetto Castelli became mathematician to the pope. One of Galileo's most ardent admirers, Monsignor Giovanni Ciampoli, was the pope's secretary of briefs and confidant.*

Galileo was convinced that the time was finally ripe for him to begin work on the book he had for many years wanted to write, on the relative merits of the Ptolemaic and Copernican world systems. This was not to be a technical tome in Latin but a work written in Italian for the educated layman, the audience Galileo felt could most help him influence Church authorities. He was appealing over the heads of the philosophers and theologians to a broader public. He called it the *Dialogue on the Ebb and Flow of the Sea*, his theory of tides being the keystone to his argument in support of Copernicus.

He set to work immediately on his return to Florence, but there is evidence that as he committed his ideas to paper he uncovered

---

*Ciampoli had written to Galileo, "It seems impossible to me that one should frequent you and not love you. There is no greater magic than the beauty of virtue and the power of eloquence: to hear you is to be convinced by your truth, and whatever I can do will always be at your service."

hitherto unsuspected weaknesses in his position that caused him temporarily to lose enthusiasm for the project. He appears to have set it aside completely between 1626 and 1629. On October 29, 1629, he wrote to his friend Elia Diodati in Paris: "I have taken up work again on the *Dialogue of the Ebb and Flow of the Sea*, which was left aside for three years, and with God's grace I have found the right line, which ought to allow me to terminate it within the winter; it will provide, I trust, a most ample confirmation of the Copernican system."[1]

Galileo took the finished manuscript to Rome in May 1630 to arrange permission to publish through his former pupil and chief censor Niccolò Riccardi. He returned to Florence shortly afterwards to avoid the summer heat in Rome, expecting to make any required revisions at his home. But a devastating outbreak of plague in Florence that left twelve thousand dead complicated communications with Rome to the point that it became necessary to print the book in Florence, causing a further delay. The book was finally released to the public in 1632 as *Dialogue on the Two Chief World Systems, Ptolemaic and Copernican.* A thousand copies were printed, a huge print run for those days.

The title had been changed at the request of the censors. There are two conflicting interpretations of the implications of this title change. Some historians argue that Galileo chose the original title because he believed the tides to be the definitive argument in favor of Copernicanism and wanted to advertise the fact that he had discovered it. The Church censors, according to this interpretation, ordered the change because they did not want to promote any theory that purported to demonstrate the truth of Copernicanism.[2] Other more charitable historians believe that the change was made for what amounts to the opposite reason—that Galileo and the censors wanted to emphasize the hypothetical character of the discussion.[3] It appears from the documentary evidence that his original intention in structuring the book had been to introduce the mystery of tidal movements, then present the Earth's diurnal and annual motion as a hypothetical explanation, and then move on to a more general discussion of Copernicanism, pro and con, as a way of gauging the merits of his theory of tides. The change of title involved a restructuring of the manuscript that placed the discussion of the tides at the end.

This restructuring reflected in turn a change in the advertised intent of the book, which may have been proposed by Galileo and accepted by the censors, or vice versa. (There is evidence to support both suppositions.) No longer a purely scientific document, it now took on an explicit political role. The book's preface, which was written jointly by Galileo and his censors, states that the book was being published to demonstrate to non-Catholics that the Church's belief in a static Earth was based on theological reasons and not ignorance, and that the scientific issues were well understood. It was intended to be, if the preface can be taken at face value, a weapon in the arsenal of the Counter-Reformation. The preface also stated that the scientific evidence seemed to favor the movement of the Earth but that it was not conclusive.

In its final form, the *Dialogue* describes four days of discussions in a palazzo in Venice among three philosophers: Salviati, who represents Galileo and the Copernican, anti-Aristotelian position; Sagredo, a supposedly neutral referee and commentator, who listens critically and accepts the arguments that seem to survive careful scrutiny (these are mostly Salviati's); and Simplicio, who rather ineffectually upholds the Aristotelian-Ptolemaic viewpoint and is frequently the butt of Salviati's barbed humor.* An amiable fellow, he seems to accept his intellectual inferiority with good grace: "I believe, verily, Sagredo . . . that I know the cause of your confusion, which, if I mistake not, rises from your understanding part, and only part, of Salviati's argument. It is true, as you suspect, that I find myself free from the like confusion; but not for that cause which you think, to wit, because I understand the whole. No, it happens quite the contrary, namely, from my not understanding anything. Confusion is in the plurality of things and not in nothing."[4]

The first day's discussion aims to destroy the Aristotelian idea that there is a fundamental difference between the terrestrial and celestial realms. The second day examines one by one the Aristotelian arguments against the Earth's movement. The third day explores the

---

*Salviati and Sagredo are named for two of Galileo's oldest and dearest friends. Simplicio was a sixth-century Aristotelian philosopher, but the name also means "simpleton" in Italian.

evidence in favor of the Earth's movement. The final day is devoted to Galileo's erroneous explanation of tidal motion. The final day presents, briefly, almost parenthetically, and in the words of the inept Simplicio, the argument Urban VIII had explicitly required to be included in any work of Galileo's touching on Copernicanism. This was the argument from God's omnipotence to the effect that no scientific model can legitimately purport to represent definitive truth. Salviati responds with the following words: "An admirable and truly angelical doctrine, which is answered with perfect agreement with that other one, in like manner divine, which gives us leave to dispute touching the constitution of the universe, but adds, withal . . . that we are not to find out the works made by His hands. . . ."[5] In context, after five hundred pages of verbose, arcane, and articulate argument, the tone seems parodic and dismissive.

Giorgio de Santillana has described the texture of the book well:

> It moves . . . on in a leisurely manner from one question into another, taking pot shots at casual objectives until we are far off the track, picks up with a "Where were we?" and comes back for a while to playing cat-and-mouse with Simplicio as a butt, but soon is off again in another direction, in full cry after some luckless lay figure who has brought up the needed asininity. Meanwhile the web of proof is being woven unobtrusively, until after a while the reader asks himself what kind of people could be blind to the evidence; what other opinion could be held except the Copernican? . . . [The characters] address each other as "Sig. Salviati," "Sig. Simplicio mio"; they quarrel and make up; they move with their feet solidly planted on the marble floors of a Venetian palace on the Canal Grande. Their forms of address are those of Italian society of the period; the scenes and interludes of action are managed by Galileo as a man of the theater who had tried his hand successfully at comedy. . . .[6]

The first copies of the *Dialogue* were in Galileo's hands in February 1632, and he proudly presented a copy to his patron, the young grand duke of Tuscany, Ferdinand II. In Florence copies were sold as fast as they could be printed, but because of the plague and quarantines, it was not until June that the book reached Rome.

He began receiving rapturous letters from fans: "The title of the work, its dedication, and preface to the reader, so excited my curiosity that before setting myself to the task of reading it . . . I could not resist skimming eagerly through . . . a part of the text, where there appear new theories and noble observations which you have rendered to such simplicity that even I, of a different occupation, am certain I will be able to understand at least some parts of it."

"The clearness with which the points are explained that seemed incomprehensible must be admired by everyone."

"You have succeeded with the public to a point at which no one else has arrived. . . . Frankly, who in Italy cared about the Copernican system? But you have given it life, and what really counts, have laid bare the breast of nature."

"Everything about it pleases me and now I can see how much stronger your argument is than that of Copernicus, though his is fundamental. . . . These novelties about old truths, new worlds, new stars, new systems, new nations, and so on, are the commencement of a new age."[7]

At the same time, the proud author began hearing rumors that the book had not been well received at the Vatican. Various historians have sought to identify both individual troublemakers and elaborate clerical conspiracies to account for what happened next, but it is doubtful whether the real story will ever be known.[8] What most agree on is that the Jesuits, at one time Galileo's staunchest defenders, had turned against him.* But even this observation obscures the fact that while Jesuit astronomers and mathematicians did indeed verify his early telescopic discoveries, and agreed with him that the Ptolemaic hypothesis had been irreparably damaged, they were from the beginning uneasy about his forays into the realm of scriptural interpretation and epistemology. Nor did most of them accept the Copernican hypothesis: Tycho's "intermediate" theory seemed to them the best bet, and it had the added advantage of not contradicting Scripture.

Once Galileo had laid his mechanistic, scientific realist cards

---

* Descartes wrote in a letter to Mersenne in February 1634, "I have allowed myself to be persuaded that the Jesuits aided in the condemnation of Galileo." Galileo himself professed to believe this to have been the case, as he made clear in a July 1634 letter to Elia Diodati.

on the table in *The Assayer*, it became clear in Jesuit circles that he had moved far beyond the point where they could continue to support him. Required by edict of their general to uphold Thomist Aristotelianism, Jesuits could not condone Galileo's contemptuous dismissal of it in favor of the new mechanistic science. And in the matters of identifying and verifying truth, and the fundamental nature of reality, they subscribed to the ideas of Cardinal Bellarmine, a senior Jesuit. Galileo's mechanistic view was increasingly seen within the order as deeply subversive of Catholic doctrine.

That being the case, the question of exactly who or what prompted the Holy Office to act becomes all but moot. The fact is that in August 1632 an order was issued withdrawing the *Dialogue* from circulation. There is some indication that the directive originated with Urban VIII himself.

Mired as he was in the hideous diplomatic complexities of the Thirty Years War, accused by Spain of being soft on Protestantism and betraying Catholic princes, Urban did not need to be seen as coddling a crypto-heretic. Just five months earlier, in the tumultuous consistory (council of cardinals presided over by the pope) of March 1632, the Spanish cardinal Gaspar Borgia, nephew of Philip IV of Spain, had publicly and violently attacked Urban. He accused him of protecting heretics and challenged him to demonstrate the same apostolic zeal as the great Counter-Reformation prelates who had preceded him. There was talk of a plot to depose the pope—worse, of poison. This was not a time for Urban to ignore or minimize what appeared to be a direct challenge to the Church's magesterium, the exclusive teaching authority bequeathed to the cardinals and the pope by Christ and the Apostles. It was, after all, the magesterium and its surrounding body of tradition and interpretation that the Protestants had chiefly objected to in their theological disputes with Rome.

And although Urban VIII was a native Tuscan, relations between Tuscany and the Vatican had been severely strained in recent years by the dispute over the succession of Urbino, a small principality in the Marches region of central Italy. At one time Church property, it had been ceded in the twelfth century to an Italian noble family. When in 1626 the last of the ruling dynasty was dying, childless, he abdicated in favor of Urban VIII, returning the territory to the papacy. Grand Duke Ferdinand II of Tuscany was outraged. He had claimed title to Urbino through blood relationships. Ferdinand threatened to seek the aid of Philip IV of Spain, but Urban had already occupied Urbino with papal troops, making the transfer of allegiance a fait accompli. Animosity simmered throughout Urban's remaining years in the papacy.

A reconstruction of events based on the documentary record suggests that Urban read the book and became incensed. Riccardi, the chief censor, was called on the carpet for having permitted its publication. Riccardi, claiming pressure from above, was able to prove that he had granted his permission only after receiving a note from the pope's confidant and secretary of briefs, Monsignor Ciampoli, who told Riccardi that Urban himself had agreed to allow its publication in an unrecorded conversation on an unspecified date. Whether this was the case is impossible to know. If Urban did give his go-ahead, it would seem probable that he did so after Ciampoli, the devoted admirer of Galileo, had assured him the book was free of offense.

Here the tale becomes complicated. Ciampoli was no longer in Rome to feel the pope's wrath after the book appeared. Following the consistory in which Cardinal Borgia had attacked him, Urban had, for safety's sake, purged his administration of pro-Spanish influence. In the process he discovered that the trusted Ciampoli had for some time been associating with friends of Cardinal Borgia. In August 1632, Urban banished Ciampoli to the governorship of the small town of Montalto, and he was never allowed to return to Rome.

But despite these peripheral pressures and irritations, the pope's immediate concern seems to have been that the book did not adequately represent the anti-realist, anti-mechanistic position he held, despite Galileo's assurances that it would.* Historians have suggested that, in addition, Urban had been persuaded to believe

---

*There are inklings of this in several letters, notably one from Filippo Magalotti to Galileo written in late June 1632. Magalotti says he has heard that among other objections, it was claimed that "two or three arguments have been omitted at the end which were invented by our Lord's Holiness himself and with which, he says, he convinced Signor Galileo of the falsity of the Copernican theory. The book having fallen into His Holiness's hands, and these arguments having been found wanting, it was necessary to remedy the oversight" (Santillana, 202).

that Galileo had slighted not just the argument but Urban himself, and that Simplicio was intended to be a caricature of the pope. Wounded vanity is thus presumed to have been a motivating factor. Most serious of all, however, was the fact that the book argued strongly in favor of Copernicanism, not just as a superior mathematical model but as an accurate description of physical reality. This was a violation of the injunction issued personally to Galileo in 1616 by Cardinal Bellarmine on behalf of the Holy Office and Paul V. The injunction, to which Galileo had agreed to conform, forbade him to "hold, defend, or teach" the Copernican theory.

At about the time the *Dialogue* was withdrawn, a document was turned up in the files of the Inquisition proceedings of 1615–16 that appeared to be a minute of the meeting between Cardinal Bellarmine and Galileo in which the prohibition was conveyed to the scientist. Its version of the injunction was even more restrictive than the one other documents show Paul V had mandated, and that Galileo had taken care to have written down in a letter from Bellarmine at the time. The newly discovered minute stated that Galileo had been enjoined not just against holding or teaching Copernicanism but against discussing it either verbally or in writing. The document is suspicious both because it contradicts several others that are clearly authentic and because it is unsigned. In the normal course of events such a minute would have been signed by Galileo and Bellarmine: without their signatures it carried little or no legal weight. Small forests have been consumed in the debate over whether the document is a forgery, how it came to be placed in the Inquisition archives, and exactly what role it played in the proceedings against Galileo. Conspiracy theorists have seen it as evidence of a plot to frame the scientist.[9] Fanatical Jesuits are usually blamed, though sometimes it is the pope himself who is held responsible for the trickery.

But there is no evidence that the missing signatures were the result of anything more sinister than a bureaucratic oversight, since the document appears authentic in every other physical respect.*

---

*A possible explanation is that the document had been prepared in anticipation of its being signed but was then not executed because the verbal understanding between Galileo and Bellarmine made it superfluous.

In the absence of corroborating evidence the fact that it goes further, and thus contradicts the documented wishes of Paul V, is impossible to account for except by speculation—and there has been an endless supply. But when all is said and done, the simple fact remains that in writing and publishing the *Dialogue*, Galileo unequivocally overstepped the bounds of both the spirit and the letter of the 1616 injunction, even in the less severe form detailed in Bellarmine's letter. In writing the book, Galileo clearly "holds" the Copernican hypothesis. The issue of the questionable minute is moot. His motive for doing so is the only imponderable.

Rather than sending the case directly to the Inquisition as he might have been expected to do, given the seriousness of the charges, Urban VIII took the unusual step of setting up a three-man commission of inquiry into the circumstances surrounding the book's clearance by Church censors and its subsequent publication. At that point the Tuscan ambassador to Rome sought an audience with the pope to press for inclusion of scholars favorable to Galileo on the inquiry panel and to ask that it hear from Galileo himself. The ambassador's report back to Ferdinand II in Florence has survived, and it makes clear that Urban was concerned not simply with the issue of Copernicus versus Ptolemy but with the deeper philosophical implications of Galileo's scientific, mechanistic views for the nature of truth and reality.

"Your Galileo," the ambassador quotes an angry pope as saying, "has ventured to meddle with things that he ought not and with the most grave and dangerous subjects that can be stirred up in these days." The pope further characterized the ideas expressed in the book as ". . . perverse in the extreme," ". . . these things which might bring religion very great prejudice, of the worst that has ever been invented," and ". . . a question of the most perverse business that could ever be handled."*

---

*We also know from Galileo's own hand that "the Jesuit fathers have insinuated in the highest quarters that my book is more execrable and injurious to the Church than the writings of Luther and Calvin" (Letter to Elia Diodati, January 15, 1633). Further evidence, if any is needed, that Urban's overriding complaint went far beyond orbital mechanics and questions of scriptural interpretation lies in the fact that the antiscientific movement within the Church grew continuously in succeeding decades, to the point that, before the century had ended,

When the ambassador asked that Galileo be consulted by the commission the pope told him that was not the way the Holy Office traditionally operated; that it examined the charges, drew up a report, and then summoned the accused to respond. "I answered," the ambassador reported, "Does it not then appear to your Holiness that Galileo should be informed beforehand of the difficulties, oppositions, and censures that are made to his work and what it is that annoys the Holy Office?" He answered violently: "The Holy Office, We are telling you, sir, does not proceed in that way and does not take that course, nor does it ever give information beforehand. It is not the custom. Besides, he knows well enough what the difficulties are, if he wants to know them; for We have discussed them with him, and he knew them all from Ourselves."

The pope ended the emotion-charged audience by insisting he had "acted with great consideration for Galileo, by having impressed upon him what he knows and by not having referred his affairs, as he ought to have done, to the Holy Office but to a specially appointed commission. . . . He concluded: 'I have used him better than he used me, for he deceived me.'"[10]

This last statement may be taken as further evidence that Galileo had explicitly agreed to incorporate the "argument from God's omnipotence" in the book.

The commission of inquiry reported in September 1632. It found that despite the 1616 injunction, the book had maintained that the Earth moves and the Sun is stationary, not as a hypothetical model but as a matter of fact; that it ascribed the phenomenon of oceanic tides to the same motion of the Earth; and that the author had, in dealing with the censors who ultimately approved the book's publication, "deceitfully" remained silent about the injunction against him.

There followed eight more specific complaints:

- The imprimatur of the Roman censor appears on the title page without appropriate permission.
- The preface [the "disclaimer" worked out in concert with

---

there was talk of banning the works of all the "new physicists," including not only Galileo but Descartes and Gassendi as well (Santillana, 205, n5).

the censors] appears printed in italic type, so as to dissociate it from the main body of text.

- The book "very frequently" deviates from a hypothetical treatment of the Earth's movement to deal with it as fact, while at the same time treating opposing argument as impossible.
- The book merely pretends to deal with its subject as undecided, whereas in fact it treats the issue as having been resolved in favor of the Earth's movement.
- The book "contemns and maltreats" authors whose views do not coincide with the author's, in particular those of opposing clergy.
- The book asserts that in matters of mathematics, human knowledge can be the equal of God's.
- The book claims as an argument in favor of Copernicanism that supporters of Ptolemy are converted to the new hypothesis, whereas the reverse is not true.
- The book ascribes tidal movement to the stability of the Sun and the motion of the Earth, which do not exist.

The commission makes no recommendations for specific action but suggests that it would be possible to revise the book to eliminate the problems: "All these things could be corrected, if it is decided that the book to which such favor should be shown is of any value."

On the basis of the report, Urban VIII forwarded the case to the Holy Office for trial. Whether he did so in anger or in sorrow is impossible to know. Galileo was formally summoned to appear in Rome the following month to face the Cardinal Inquisitors.

A series of attempts were made by Galileo and the Tuscan government to prevent his having to place himself in the hands of authorities in Rome. It was first suggested that he be sent a list of questions to which he could reply in writing. Then it was suggested that the trial be turned over to the Tuscan Inquisition in Florence. When neither of these proposals met with approval, an affidavit signed by three Florentine doctors was sent to Rome, attesting that Galileo, at sixty-eight, was too old and infirm to undertake the two-hundred-mile journey from Florence to Rome.

Galileo himself wrote a sad, pleading letter to the pope's nephew, Cardinal Francesco Barberini, one of the most powerful members of the Roman curia:

> Both my friends and I foresaw that my recently published *Dialogue* would find detractors. Of this we were assured by the fate of other works of mine previously printed, and because it seems that this generally happens with doctrines which distinctly depart from common and inveterate opinions. But that the hatred of some men against me and my writings . . . should have had the power to convince the most holy minds of the superiors that this book of mine is unworthy of publication was truly unexpected. . . . Whenever I think of it, the fruits of all my studies and labors over so many years, which had in the past brought my name to the ears of men of letters with no little fame, are now converted into grave blemishes on my reputation, giving a foothold to my enemies that they may rise up against my friends . . . and say that finally I have deserved to be ordered before the tribunal of the Holy Office, a thing that happens only because of the most grave delinquencies. This affects me in such a way that it makes me detest all the time I have spent in those studies, by means of which I hoped and aspired to separate myself somewhat from the trite and popular thinking of scholars; and by making me regret that I have exposed to the world a part of my compositions, it causes me to wish to suppress and condemn to the flames those which remain in my hands, entirely satiating the umbrage of my enemies, to whom my thoughts are so troublesome.[11]

At the same time, friends suggested to Galileo that he should flee to Venice, or to Holland, where he would be out of the reach of the Inquisition. Whether he seriously considered these options is not known. Given his age, his essentially pious nature, and his strong attachment to his eldest daughter cloistered near his villa at Arcetri, it seems doubtful.

In late December the Inquisition sent Galileo an ultimatum: either he appear in Rome forthwith or be placed under arrest by a commissioner. He would be examined by a physician at that time, and if travel proved an authentic threat to his life, the trip would be postponed but only until he had recovered, when he would have to make the trip in irons.

Having run out of room for negotiation, Grand Duke Ferdinand II finally heeded the persistent entreaties of his secretary of state and advised Galileo to obey the summons. He put the royal litter and its associated retinue at the scientist's disposal, and ordered the Tuscan ambassador to Rome to do everything in his power to assist in Galileo's defense.

Five days before his departure, Galileo wrote a long letter to Elia Diodati in Paris, in which he set out his position vis-à-vis the Church as if making a last testament of his beliefs and placing it in the hands of a friend who would be in a position to publish it. In his characteristically trenchant prose, he makes his position very clear, in a passage that takes the form of a response to anti-Copernican tracts by an obscure author named Froidmont:

> For if I were to ask Froidmont, who it is that made the Sun, the Moon, the Earth, and the stars, and ordained their order and motions, I believe he would answer, "They are the creations of God." If asked who inspired Holy Scripture, I know he would answer, "The Holy Spirit," which means God likewise. The world is therefore the work and the Scriptures are the word of the same God. . . . Nothing ever changes in Nature to accommodate itself to the comprehension or notions of men. But if it be so, why, in our search for knowledge of the various parts of the universe, should we begin rather with the words than with the works of God? Is the work less noble or less excellent than the word? If Froidmont or anyone else had settled that the opinion that the Earth moves is a heresy, and if afterward demonstration, observation, and necessary concatenation would prove that it does move, into what embarrassment he would have brought himself and the holy Church.[12]

The essential error in his position remained undiscovered by Galileo even at this late date. It lies hidden, in this case, in the sentence "Nothing ever changes in Nature to accommodate itself to the comprehension or notions of men." The implied continuation of the argument is, scriptural interpretation, on the other hand, is subject to human foible. Therefore science's descriptions of nature must have a superior claim to truth. This was the "most perverse business" that so disturbed Urban VIII.

To imagine that humans could see the underlying reality of nature was to fall into a dangerous error. God and nature could be known, but only through the knowledge offered to humankind by way of received wisdom, or revelation, or spiritual insight. Although it could sometimes be partially confirmed by reason, this was knowledge that ultimately had to be accepted on faith. But it was also a knowledge that was capable of illuminating those aspects of understanding that *were* accessible to human reason. Faith and reason could thus be mutually reinforcing.

A thousand years of Scholastic inquiry founded on a further millennium of Classical wisdom had taught the scholars of the Church these essential truths, and informed them of their implications.*

Nevertheless, as Galileo finally went to trial in April of 1633, after six long and anxious weeks of waiting to be summoned from his palatial surroundings in the Villa Medici, that hard-won knowledge was on the brink of being supplanted, virtually overnight, by his simplifying, mechanistic science.

---

*Of course, this knowledge was by no means held exclusively in the Roman Catholic Church. But it is the Church that is here the focus of interest.

TWENTY

An Enigmatic Pope • A Mass in St. Peter's
A Chance Encounter

Shortly after my visit to Santa Maria Maggiore and its Pauline Chapel with Berkowitz, I vacated my comfortable room at the Hassler Hotel on the Pincean Hill to set up camp in the much older and more Spartan surroundings of the Hotel Columbus, just a stone's throw from St. Peter's and the Vatican.

My third-floor quarters overlooked the Via della Conciliazione, the grandiose boulevard Mussolini ordained to celebrate the signing of the 1929 Lateran Treaty, establishing Roman Catholicism as Italy's state religion and Vatican City as an independent sovereign state. This was the entente that formally ended the diplomatic standoff between the Vatican and Rome over the confiscation of Church lands and properties that had accompanied Italy's unification. The Hotel Columbus, built as a private residence by Cardinal Dominico della Rovere in 1490, was one of only a handful of medieval and Renaissance structures to survive Mussolini's determination to raze several blocks of buildings in order to create the sort of imposing vista and broad, sterile artery that autocrats admire. It was finished in 1937, complete with two dozen travertine obelisks supporting huge wrought-iron street lamps goose-stepping down the curbs. It remains one of the few places in Rome that is unfriendly to pedestrians, where the scale feels not just superhuman, but inhuman.* Many of those who watched the project unfold, and who remembered the old approach to St. Peter's, down narrow medieval streets lined with shops and artisans' studios, and the breathtaking surprise of stepping into the sunlit vastness of St. Peter's Square with Bernini's enfolding columns, its fountains, and the ancient Egyptian obelisk, felt a great deal had been lost in the creation of a streetscape best enjoyed by speeding motorcade or from the air.

Much of the wonderful late Renaissance and Baroque art and architecture of Rome is a tribute to the tremendous energy and resolve of the Counter-Reformation. It began with the reign of Pope Paul III (1534–49) and concluded under Urban VIII. The trial of Galileo may be thought of as one of the final manifestations of its zeal. By the time of the election of Maffeo Barberini to the Throne of St. Peter in 1623, the sternness of the face of the Church had relaxed somewhat in the light of the Counter-Reformation's successes. The spread of Protestantism had been checked, and the Church of Rome was expanding its influence both in Europe and the New World. Ottoman Turkey had been soundly defeated by the allied forces of the Vatican, Venice, and Spain at the great naval battle of Lepanto. Henry IV, the first Bourbon king of France, had been converted to Catholicism before his accession to the throne. Rome was resplendent in gleaming new churches and ecclesiastical buildings of every description, not to mention the new aqueducts, fountains, and public squares.

Nevertheless, the times remained uncertain enough that Urban VIII thought it prudent to build a new encircling wall around the

---

*Another is the area surrounding the preposterous monument to King Victor Emmanuel II, which attempts to place the unification of Italy under this Sardinian monarch on the same scale as the achievements of Imperial Rome. Its garishness and bloated scale make it the most prominent feature in Rome's otherwise enchanting earth-toned skyline. Begun in 1885 and completed in 1911, architect Giuseppe Sacconi's gleaming white Altar of the Fatherland with its elephantine statuary destroyed a charming Renaissance square in the building. It has a creepy proto-Fascist feel about it, which is duplicated in the king's tomb in an alcove of the Pantheon. It is in reality a monument to the pathetic vanity of nineteenth- and twentieth-century European autocrats. Rome would be well served if it were to be torn down to expose the ancient ruins that are known to lie beneath it.

already heavily fortified redoubt of Castel Sant'Angelo, and to see that the papal apartments there were suitably decorated and the larders amply stocked. He built Castelfranco on the northern frontier of the papal territory and improved fortifications for the port at Civitavecchia. Although the Thirty Years War of 1618 to 1648 had by his coronation come to involve virtually all of Europe *except* Italy, it was no time to neglect defenses. The war was being fought just over the Alps in Germany, and the ravening armies of mercenaries that swept back and forth across that benighted land were scarcely less disciplined than those that had ravaged Rome a century earlier. The suffering and privation among the civilian population was unprecedented.

One of the great watersheds of European history, the Thirty Years War had begun as a war of religion when Emperor Ferdinand II of the Holy Roman Empire tried to suppress the Lutherans and Protestant princes allied against him. But it soon evolved into a dynastic struggle between the (Catholic) Habsburgs of the Holy Roman Empire and Spain on the one hand and the Bourbons of (Catholic) France and their (Protestant) Swedish allies on the other. The Swedish intervention led to a series of spectacular successes in the field, which provoked the Spanish and Imperial ambassadors in Rome to accuse Urban VIII of assisting the Protestant cause. Not only had he refused to declare the conflict a holy war and failed to provide any significant material support to the Catholic cause, they said, but he was actively conniving with the Protestants against the Habsburgs.

By the time the war ended, the world had become almost unrecognizable. The Holy Roman Empire was dead beyond hope of resurrection, as was the ideal of Christendom unified in its religious beliefs and standards. Roman Catholicism had lost its thousand-year-old position as the main inspiration and adhesive of European civilization, and religion had become ancillary to national politics. Increasingly, in the minds of a people hardened by war and privation, the new, mechanistic philosophy was ascendant. The Age of Reason was lurking just around the corner. Urban VIII's new wall around the Castel Sant'Angelo was as irrelevant as Aurelian's great wall around Rome had been.

More and more, as I haunted the precincts of the Vatican, I wondered what sort of man Urban VIII had been. The history books describe him as accomplished in the arts, a good poet, and a passable musician, a man of taste. He was well educated in philosophy and science and trained in the arts of diplomacy. He was perceptive enough to understand the threat to the role of religion posed by Galileo's philosophy and yet appreciative of its power to shed light on the mysteries of creation. He was simultaneously the last pope of the ancien régime and the first pope of the modern era. On this much, most historians agreed.

But a mystery remained: why had he turned with such harshness on Galileo? Was it something in his character that caused him to switch from being the scientist's admirer and protector to his prosecutor and jailer? Or was he merely a victim of his position in the power structure of his time, as Brecht suggests in his dramatically compelling but historically flawed *Life of Galileo*? In the play, the scene in which the pope is being dressed in his robes of office while a malevolent Inquisitor badgers him, against his better judgment, into prosecuting Galileo for promoting "filthy skepticism" is meant to suggest that social being determines thought and that in this case official policy has replaced the pope's private views. Urban is cast as a hypocrite.*

It was a perfect late-spring morning when I slipped my opera glasses into my pocket, left my hotel, and walked up the Via della Conciliazione and across St. Peter's Square to the Basilica, to take a close look at one of its many treasures, the tomb of Urban VIII by Bernini. There was the usual crush of tourists and pilgrims from around the world, but Vatican security is efficient and friendly and I was soon through the airport-style metal detectors and beyond the eagle-eyed policemen checking camera cases and purses, and mounting the marble steps to the portico. As always, the sheer scale of the building impressed me, all the more so because even from as

*The difficulty with Brecht's interpretation is that it sees the Church as maintaining a cynically anti-intellectual stance in its dealing with Galileo in order to preserve its power and influence, which, Brecht says, is derived from ignorance and unreasoning faith. The best that can be said about this argument is that it is highly anachronistic: as a Marxist, Brecht looks back on the Galileo affair with more than considerable hindsight.

short a distance away as my hotel window, it did not seem inordinately large. It is so well proportioned that only the sight of people moving, ant-like, up the steps and through the giant columns supporting the portico roof indicates its true size. It replaces an earlier church built over the tomb of St. Peter by Emperor Constantine in 324, and was 120 years in the making, spanning the careers of twenty popes and ten architects, of whom Michelangelo and Bernini are the best known. Urban VIII formally consecrated it in 1633, the year of Galileo's trial.

It is the largest church in Christendom and its many splendors are so well known that there is no point in describing them here—except perhaps to point out that the subjective experience of a Baroque cathedral is entirely different from that of the soaring, austere, Gothic style typified by Chartres or Notre Dame de Paris. In the wealth and exuberance of their decoration, Baroque churches are unabashed, even prideful celebrations of the arts of man. Where earlier artists had sought to convey permanence and stability, regularity and predictability, there was now a motion and a restlessness, a straining at the bonds of convention. Columns twist and swirl, the wind tugs at billowing garments, pictures spill out of their framing, sculpture is combined with painting and stained glass in startlingly ambitious attempts to represent the transcendent. The psychological gap between God and humanity, one senses, has narrowed in the four hundred or so years that separate the two architectural epochs. In that time the architectural and artistic vocabulary has shifted slightly but unmistakably from the mysterious toward the material, and more obviously from the cool and allegorical to the hot and literal.*

There was a Mass being said that morning—there were many special services each week throughout the Jubilee year—and there

---

*The art historian E. H. Gombrich provides salient advice on understanding the Baroque in ecclesiastical architecture in *The Story of Art*: "It is not so much the details that matter in these interiors as the general effect of the whole. We cannot hope to understand them, or to judge them correctly, unless we visualize them as the framework for the splendid ritual of the Roman Church, unless we have seen them during High Mass, when the candles are alight on the altar, when the smell of incense fills the nave, and when the sound of the organ and the choir transports us into a different world" (Phaidon Press, 1991, 345).

were perhaps five thousand people visiting the Basilica when I arrived. But it was by no means crowded; it will reputedly hold sixty thousand, though I would not want to be among them. I walked through the Holy Door, past the many treasures I had become familiar with over the past few days, past Michelangelo's flawless *Pietà* (behind glass since a maniac attacked it with a hammer thirty years ago), past the astonishing bronze baldaquin over St. Peter's tomb, on to the apse where Bernini enclosed St. Peter's episcopal chair (already encased in ivory at the behest of Charles the Bold in 875) in a massive, sculpted bronze throne supported by four great Doctors of the Church and surmounted by an amazing gilded stucco "gloria" with swirling clouds and a host of cherubs and a golden central window through which the sun pours with otherworldly effect in the afternoon. In alcoves to the left and right of this focal point are two more Bernini masterworks. On the left, his monument to Pope Paul III and on the right to Urban VIII. While the service continued and the sound of a choir filled the immense space beneath Michelangelo's dome, I maneuvered myself as close to Urban's tomb as I could and took out my little binoculars to examine his face, for I knew that Bernini could be counted on for accuracy.

There is a famous bust of France's Louis XIV in the Musée de Versailles that captures the Sun King as if in a videographer's freeze frame, just as he is about to speak. It is a trademark Bernini work, capturing the moment, and in that moment the enduring essence of his subject's character. It is an approach that practically defines Baroque art. Happily, we have a detailed record of how Bernini went about creating this portrait, in the diary of the king's steward, a man who served as Bernini's translator and liaison with the court during the artist's sojourn in France late in his career. Having personally selected a suitable block of marble, the diary tells us, Bernini set about creating a clay model, in which he developed the basic form of the piece, with its flowing wig, armor, and fabric and its pedestal. He was, in this part of the process, not much interested in the exact details of the face but sought to achieve an overall impression of majesty appropriate to his subject. At the same time he undertook, as he said, "to steep myself in, and imbue myself with, the King's features."[1] He observed Louis on the tennis court, in meetings of his executive council, hunting on horseback, all the while making sketch after sketch. "If a man stands still and is immobile," he said, "he is never as much like himself as when he moves about. His movements reveal all those personal qualities that are his and his alone."[2] At the end of three weeks, he was ready to attack the marble. He never again referred to either his preliminary clay models or his sketches. "I don't want to copy myself," the royal steward records him saying. The bust took forty days to complete; Louis sat for the artist on fourteen of those days.

Assuming that something like the same painstaking striving for realism went into the sculpting of the Urban VIII monument, I hoped to get some sense of the man by carefully examining the statue. However, because of the portable railings set up to provide a buffer between the visiting public and the congregation for the Mass—a gathering that seemed to include a great number of senior clerics of various degrees—I was unable to stand directly in front of the statue. Nevertheless, I could get a reasonably good view. Like everything else in that place it was enormous, in keeping with the scale of the building. What I saw was the robed and mitered pope raising his hand in blessing, sitting on a throne above his own sarcophagus, around which are grouped statues of Justice, Charity, and Death—who is inscribing the pontiff's name.

I raised my glasses and discovered a face that seemed to me remarkably serene, especially when compared with the worried, stern, angry, even anguished expressions in other, earlier papal portraits I had seen in Rome. The forehead is high and clear, the nose aquiline, the eyes large and heavy-lidded. The lips are obscured somewhat by his moustache and goatee, but there is a hint of a smile. Was there a touch of vanity in that smile? I could not tell whether I was seeing it or projecting it onto what I was seeing. So many accounts speak of Urban's personal vanity, giving as evidence the fact that he commissioned his own funerary sculpture and had it placed prominently in St. Peter's. A humbler man might have waited for his successors to judge his importance. Then again, that brand of modesty was so rare among the ruling classes of the time that it likely would have been interpreted as weakness, which was the last thing a pope of that dangerous era could afford to display.

As I was recording my impressions in my notebook, the words

of the sermon under way behind me began to intrude, and I paused to listen more closely. Though it was being delivered first in Italian and then French, I gleaned that the subject was science. I fished through my shoulder bag for the Jubilee program I had picked up as I entered the Basilica and checked the date. There were special Jubilee days listed for many groups and professions—for teachers, for migrant workers, for journalists, for the clergy, for students—and today was the Jubilee for scientists. Considering my reason for being there, it seemed a remarkable piece of serendipity. All the more so when the speaker switched to English briefly to state that scientific research could not be artificially restricted without being made "impoverished, in fact, a waste of time." I scribbled notes furiously. Scientific research ought to be done, he went on, in recognition of the ultimate mystery of God and in the service of the betterment of man. In other words, he said, science, in order to be truly and genuinely successful, must be informed by faith. The scientist who works in this way, he concluded before switching back to Italian, will have "a different light, a deeper genius, a clearer awareness, and a higher calling. People imbued with God's wisdom, and enlightened by this inner strength, surpass the boundaries of science, scrutinize the powers of spirits and see and understand all that is hidden: there is something of the prophet in scientists! Scientists who can do their work with love are invested with a prophetic charisma for the third millennium."

I was reading over my notes of these last words, trying to digest them, when a sotto voce greeting from a familiar voice caused me to look up. It was Berkowitz.

"Good grief! Fancy finding you here in this mob," I said. It was one of those coincidences that can make fact seem stranger than fiction.

"It wasn't hard to spot you," he said. "You were the only one staring at statues with binoculars while the sermon was going on!"

"Did you hear that?" I nodded to where the cardinal was wrapping up his homily with a prayer.

"Most of it. What did he mean by 'prophetic charisma'?"

"Wow. Yes! That is quite a remarkable phrase, isn't it. As far as I know, *charisma* means an extraordinary power, for instance the ability to perform miracles, that's granted to a Christian by the Holy Spirit."

"So he's saying—"

"He's saying that scientists who pursue their research inspired by faith, or 'in the hand of God,' as he says, can perform miracles of insight."

"In other words, a scientist can be a prophet."

"I don't know much about modern theology, but that's what it sounds like, doesn't it? But only a certain kind of scientist—one who recognizes and accepts the existence of a higher level of reality. He has to first accept that there is knowledge that is inaccessible to straight science, and then he'll be able to 'see and understand all that is hidden.' That's what I make of it, anyway."

"Scientists as prophets! Bellarmine must be spinning in his grave!"

"Why do you say that? I don't think so. I don't think Bellarmine would have any basic disagreement with the cardinal here. It's really just an update on the idea of 'learned ignorance,' which is about five hundred years old."

"Learned ignorance," Berkowitz chuckled. "What kind of gobbledygook is that?"

"I'm headed for the Capitoline, if you'd care to join me. Some more statues to examine. We can talk on the way."

"Thanks, but no thanks. I'm overdue for lunch and a siesta. Look forward to tonight, though."

"You're still on for dinner with Sister Celeste?"

"Is the pope Catholic?"

I winced. "We'll talk about it then."

TWENTY-ONE

A Remarkable City • History's Parochialism

An Uncomplicated Man

The Motives of Pontiffs

I left St. Peter's feeling this was a lucky day for me and headed in the direction of the Capitoline museums in the center of the city. There, I knew, I could find another statue of Urban VIII, also by Bernini. I crossed the Tiber on the venerable Ponte St. Angelo, although it was slightly out of my way, just to see Bernini's eight wind-blown angels once again. From there I threaded my way through the endlessly fascinating labyrinth of narrow streets and tiny piazzas, poking my nose into antique shops and used book stores and bakeries and wine and oil cellars to the Piazza Navona. There I paused to admire the glorious Fountain of the Rivers by the ubiquitous Bernini before continuing my circuitous path east past the Pantheon, down Via della Minerva, past the storied Santa Maria sopre Minerva, south from there across the noisy, smelly, and congested Corso Emmanuel II and into another rabbit's warren of medieval streets and alleyways where I happily resumed my browsing. I was soon at the Piazza Aracoeli, from where it is just a short, death-defying sprint across the shooting gallery known as Via dei Teatro di Marcello to Michelangelo's stately staircase mounting to the top of the Capitoline, my destination. The staircase, and the Piazza del Campidoglio with its embracing palaces, the giant antique Roman statues of Castor and Pollux, and the huge equestrian statue of Emperor Marcus Aurelius at its center, were laid out by Michelangelo at the request of his patron Pope Paul III, who wanted a suitably impressive venue to welcome his adversary Emperor Charles V.

As I paused to catch my breath at the top of the staircase, I could not help thinking once again what a remarkable city this is. The palace facing me across the square, the thirteenth-century Senator's Palace, is now the Rome City Hall. The "new" facade is by Michelangelo, as are the two adjacent palaces, now museums. In the sixth century B.C. this was the site of an Etruscan temple to Jupiter, that mysterious civilization's most sacred sanctuary. The Campidoglio, for which the square is named, is one of the seven hills on which Rome was built. Its two summits are called Capitoline and Arx (Citadel), and the square is in the dip between them. In Roman times, it provided an entrance to the Forum (the sprawling ruins of which stretch out beyond the square), and held temples to Jupiter, Juno, and Minerva, the Capitoline Triad. The place is saturated with enough history and myth to fill a volume on its own, from the eighth-century B.C. Rape of the Sabine Women to the story of the Geese of the Capitol who warned of a third-century B.C. attack by the Gauls. It was on this spot, recalled Edward Gibbon, that "on the fifteenth of October [1764] in the gloom of the evening, as I sat musing . . . while the barefoot friars were chanting their litanies in the temple of Jupiter, that I conceived the first thought of my history . . . the decline and fall of the Roman Empire."

It seemed impossible that anyone fortunate enough to experience the layers of living history of this place and its incredible artistic richness could come away without being changed. The untraveled and ill-educated can easily be convinced that their little corner of the planet is the best of all possible worlds. It is not a far step from there to a jingoism that denigrates the outsider, as a means of maintaining the psychologically important myth of local superiority. History, too, has its parochialism, just like geography. Briefly stated, it amounts to this: past = bad; present = better; future = best. Most of us today believe we live in the best of all times, on the leading edge of social evolution and cultural development. We have survived a nasty and brutish past to arrive at a comfortable, prosperous present and can look forward to a utopian future. We see answers to our increasingly burdensome human problems exclusively in pursuing the same goals that caused the trouble in the first place—in more "progress." How, indeed, can the past offer solutions, when it is by definition inferior in every respect? A few well-spent weeks in Rome, or in the medieval towns of France, or in the history-drenched lands of Greece—a trip back to times when civilization's priorities were radically different from ours, cannot help but unsettle that belief, and awaken us to a more complete spectrum of possibilities. Goethe said his life began when he first visited Rome: I think this may be what he meant.

I bought a ticket to the museums and soon was standing before Urban VIII once again, merely twice life-size this time, in the Horarii and Curatii Room of the Museo del Palazzo dei Conservatori. Once again he is seated, but now both arms are outstretched, as if welcoming a friend. Again, the look of intelligent good humor is remarkable. He is clearly about to speak, perhaps to utter some ecclesiastical witticism to a scholarly guest. I could see no hint of the narcissism I suspected in the Vatican piece. At the other end of the room a bronze statue of his successor, Innocent X, competes for attention. The work is one of the sculptor Allisandro Algardi's masterpieces; Algardi himself is universally regarded as second only to Bernini as a Baroque sculptor. Innocent's face, as I examine it with the aid of my binoculars, was a study in determination and hauteur and, to my eyes, tinged with cruelty. He is clearly not a man to be trifled with. I walked back to look up at Urban and wondered once again what to make of this man. Could it be that Bernini's judgment had been clouded by his friendship for his powerful benefactor? On the other hand, I knew that the artist, from very early on in his career, was able to charge enormous sums for his work and was a very wealthy man—immune, one would think, to economic pressure to pander to his patrons' vanities. If only there were some other perspective, by an artist other than Bernini, I thought.

And this being my lucky day, instead of doing any of a thousand and one other things I could have done, I wandered upstairs to the Pinacoteca, the gallery of paintings hanging in a series of interconnecting rooms, strolling past Titians and Rubenses and Volets and Caravaggios—already, after so short a time in Rome, taking the city's staggering artistic riches for granted. And as I walked through the door into the last room in the gallery, a larger space that had been set up with chairs and a podium for lectures, there on the wall in front of me was a portrait of a scarlet-caped cardinal whom I knew immediately to be Maffeo Barberini—Urban VIII. The painting, in oil, was about three by five feet and the card on the wall beside it informed me it had been made in about 1627 by Pietro da Cortona, a contemporary of Bernini ranked as one of the outstanding Baroque artists.* The man it portrayed was essentially identical to the one depicted in the Bernini sculptures—the same Mona Lisa smile, the same intelligent visage, the same aura of amiability. It was the face, it seemed to me, of an uncomplicated man.

I was led to reflect, as I had many times in studying the history surrounding the Galileo affair, on how difficult it can be to understand the motives behind the actions of popes. Since the eighteenth century, academics in the liberal arts, like their colleagues in the sciences, have increasingly tended to regard religious motivation with suspicion or frank disbelief, if not in their private lives, certainly in their public writings. For a scholar to do otherwise is to risk being politely shunted to the sidelines of academia. While it remains "academically correct" to accept at face value the religious motivation of a Francis of Assisi or a Catherine of Siena, it is not safe to do the same of popes, especially those whose reigns have had a clear impact on the course of European history. The loftiest of religious motivations are invariably reduced by modern interpreters to the status of mundane scheming for secular power. Of course, many a pope has engaged in exactly that practice, on behalf of himself, his family, and the papacy.

But it seems foolhardy to ignore the fact that the Church of Rome had, and continues to have, as its major goal and preoccu-

---

*His most famous work is the ceiling fresco in the main salon of the Barberini Palace, which is just off the Via Veneto, near the Intersection of the Four Fountains. Intended as a glorification of the reign of Urban VIII (who built the palace), it is entitled *Allegory of Divine Providence*. It is an illusionist work in which the ceiling is divided into sections by a painted framework imitating architecture and sculpture. Scores of human and angelic figures soar both above and below the frame, creating both a truly remarkable 3-D effect and the illusion of unbounded space in the sky above. "Here," a historian has noted, "the baroque style reaches a thunderous climax."

pation the teaching of the Gospel and the salvation of souls. The actions of all but the most venal of men occupying the Throne of St. Peter can only be sensibly interpreted if this essentially irrational fact is taken into account, along with the equally unscientific idea that popes are considered by the Church, and presumably by themselves, to be the direct spiritual descendants of the apostle Peter, who was personally entrusted by Jesus Christ with the building of the Christian communion through the Church. These facts, however awkward to concede in the context of modern scientific scholarship, *must* be acknowledged, if only to realize that the Church tends to take a much longer view of history than do politicians or even monarchs. The Church was well over a thousand years old before the first of the European nation-states was born, indeed, before their national languages appeared.

Popes are often criticized for acting in the interest of the Church, particularly if it is at the expense of some national interest. Thus, the historian Friedrich Heer, in his study of the Holy Roman Empire,[1] severely criticizes Urban VIII for his policy of "dissembling" in the face of the emperor's attempts to come to terms with the Protestant warlords for the sake of peace and Christian unity. In Heer's mind, Urban's failure to wholeheartedly support the Catholic League, set up in the Thirty Years War to oppose the Protestants, is evidence of indecisiveness and unscrupulousness. "The Pope," he reports, "had only one real political interest at heart, the Papal State and the Barberini family."

This seems to me a fatuous statement. But it is by no means atypical of historical writing, both academic and popular, where it concerns the papacy.* The facts, even as outlined by Heer, support the more obvious hypothesis that Urban VIII was acting throughout the Thirty Years War in what he and his advisers saw as the best long-range interests of the Church. It is clear that he understood

---

*In the same category, to quote just one more example, is the MIT professor Giorgio de Santillana's characterization of Paul V as ". . . not an open mind, nor much of any kind of mind" (in *The Crime of Galileo*, 118). I strongly suspect that had he had the opportunity to meet and debate with Paul V, or indeed any other late Renaissance pope, Professor Santillana might have been moved to concede his own limitations in the face of their enormous intellectual accomplishments. These were, after all, Renaissance men, exceptional even for their times.

that the war was being fought for dynastic, and not religious, reasons. And it must never be forgotten that Urban and his many advisers believed, more or less fervently, that what was best for the Church was best for the eternal salvation of human souls, which is, to repeat, the continuing aim of this most ancient of Western institutions. Whether or not one believes in the existence of souls, and whether or not one subscribes to the beliefs of Christianity or the dogmas of the Church of Rome, the *Church's* belief remains an inescapable fact of history.

This difficulty in the writing of history is already evident in Galileo's time, as the Oxford historian G. N. Clark points out in *The Seventeenth Century*:

> When the Reformation and the Counter-Reformation brought religion and politics into such close relations that neither could be chronicled without the other, historians took their part in the confessional strife and every period of Christian history was written or rewritten by conflicting apologists. . . . Those who were openly partisans insisted most on the worldly intrigues and ambitions of their opponents in their own or earlier times, whether these were popes or heresiarchs. Impartiality was to be found only in those writers who neglected the religious motives of both sides alike. Thus there is superficiality and a poverty of human sympathy in the best authors of the beginning [of the seventeenth century].[2]

Clark uses the Venetian patriot Paolo Sarpi's scrupulously detailed account of the Council of Trent as a prime example of this kind of bloodless, misleading history: "His [Sarpi's] long history of the Council of Trent, first published in London in 1619, was long held in high esteem for its insight and for its systematic use of original authorities. Yet it treats the council and the movement which led to it as nothing more than a series of moves and counter-moves in a game of domination: there is scarcely a hint in it that the council was the decisive phase in a great reformation of the Roman Church from within."[3]

Historians need to evaluate the actions of popes, at least to some degree, in the context of the outcomes of those actions relative to the Church's historic goals of the spreading of the Gospel and saving human souls. There is obviously a great difficulty in knowing *to what extent* the pursuit of these goals, as opposed to other, more venal or secular aims, has influenced any given decision. But it is almost always a mistake to ignore the religious motivation entirely. The biggest papal decisions, of course, have usually come down to a choice among several more or less unattractive options, as was the case for Innocent III when he launched his terrible Crusade against the Cathars of southern France in the thirteenth century. The best popes spent a good deal of their time weighing these sorts of moral dilemmas, drawing on the Church's long history of thought and experience and asking for divine guidance through prayer, although of course not always coming up with the right answer.

Urban VIII's papacy was one of extraordinary challenges, coinciding as it did with one of the defining upheavals of European history, the Thirty Years War. It would be an unfair commentator who would not concede that he had acquitted himself reasonably well in the circumstances, and that the Church was at least as strong when he died as it had been when he ascended to the throne.

But it was never to be so strong again, and that, too, is part of this affable, accomplished pope's legacy. To Urban VIII fell the unlucky duty of finding a way to deal with Galileo and the protean philosophy of scientific realism he represented.

TWENTY-TWO

A Dialogue in Vatican City

Galileo's Rehabilitation • "Learned Ignorance"

The Puzzle of the Abjuration

"Who was that French-speaking cardinal who delivered the homily this morning?" Berkowitz demanded over dinner.

"No idea," I said. "But it was a pretty good summing-up of the Church's current position vis-à-vis science, I thought."

"You were at St. Peter's?" Sister Celeste asked.

I nodded. "Both of us, as it happened."

"You were fortunate! I had hoped to attend myself, but I had to work. You must be referring to Cardinal Paul Poupard, Mr. Berkowitz."

"That name rings a bell," I said. "Didn't he have something to do with the Galileo Commission?"

"Indeed he did. He was in charge of it. He is the president of the Pontifical Council for Culture, which seven or eight years ago took over the responsibilities of the Pontifical Council for Dialogue with Nonbelievers." She paused to acknowledge Berkowitz's sardonic chuckle. "An awkward name, I agree. But its purpose was and is to provide liaison with other cultures, not so much in the ethnic as in the philosophical sense. For example, Cardinal Poupard has attended the Venice Film Festival on behalf of the Vatican to lobby for—or perhaps I should say to bear witness to—spiritual values in cinema."

"Quite a challenge," Berkowitz said, laughing.

"Indeed. And so he was the one asked to run the Galileo Commission. He is also very close to the Holy Father. He often represents him at dedications and so on. You are aware of the Galileo Commission?"

"Remind me," said Berkowitz.

Sister Celeste looked to me for an answer, but I suggested she should carry on. I enjoyed listening to her impeccable Italian-flavored English.

"It is John Paul II's position," she said, "that the Galileo affair was an unfortunate clash of cultures that could well have been avoided. Because it has tarred the Church with a reputation for intellectual intolerance, and because he feared it could easily happen again, he wanted the issues brought out in the open and cleared up, at least as far as the Church goes, once and for all. So in 1981, he set up the study group within the Pontifical Academy of Sciences, which commissioned research, began a large program of publishing, and finally reported back in 1992. I myself had a modest role in preparing the report. Cardinal Poupard was the commission chairman."

"I remember reading about that in the papers," said Berkowitz. "'*Pope forgives Galileo, admits Earth revolves around Sun*,' or something like that."

"Yes, there were headlines like that, unfortunately."

I could see Berkowitz's remark had touched a nerve. Her cheeks flushed.

"It really is a mystery to me how the media can assume such stupidity on the part of the Church! Cardinal Poupard has doctorates in both history and theology. The people on the commission were leading historians, scientists, theologians, and philosophers. Do the media really think that such an eminent group would make such a vacuous and foolish report as that? Do they think the Holy Father himself is a stupid man? I really cannot grasp it!"

Berkowitz was a little taken aback by her outburst. "That's only what I read. I'm sure—"

"You can be sure, Mr. Berkowitz, that the commission delved

very deeply into the question of faith and reason, science and religion, going all the way back to St. Augustine and beyond. There is, after all, a legacy of two thousand years of intense thought by some of history's greatest minds to draw on. And they looked very closely at the mistakes the Church made in dealing with Galileo. You can be sure of that!"

"Not to change the subject or anything, but can we get back to that 'learned ignorance' thing you were talking about this morning?" Berkowitz asked.

"You're not changing the subject," I said. "It's all the same topic. It's an idea of Nicholas of Cusa, a fifteenth-century cardinal, the bishop of Brixen. One of these bright people Sister Celeste was talking about. It takes St. Augustine's *pia confessio ignorantiae*, the humble confession of ignorance, and extends its meaning. Augustine had in mind the admission of ignorance as a point of going over into faith. The Cusan puts a more positive spin on it. He sees the confession of ignorance as the first, necessary step to exploring that ignorance. Ignorance was to him an opportunity to expand knowledge. It was terra incognita, waiting to be explored. Ironically, it was the rationalist Rousseau who best summed up the idea, in a letter to, of all people, Voltaire, arch-enemy of religion. He says that what we do not know harms us less than what we think we know, but don't. 'If we had not pretended to know that the earth does not turn, then no one would have punished Galileo for having said that it turns.'"[1]

"A perfectly sensible comment," Berkowitz interjected.

Sister Celeste picked up the point. "So it might seem. Except that Rousseau and his friend Voltaire, like the rest of the Rationalists, failed to see that if the Church had committed this error, so had Galileo. And the reason they did not see this was that they, too, were caught in the same trap of failing to see their own ignorance and being prepared to learn from it."

"Of course they saw their ignorance. That's what science is all about. Filling in the blanks in knowledge," Berkowitz asserted.

"Yes, but only certain kinds of knowledge. The rest, they ignore. Rather than confront their ignorance of it, they insist it does not exist."

"I think we've covered this territory."

"Apparently not thoroughly enough, Mr. Berkowitz. Let me

pour you a glass of wine. I want to tell you a story."

Berkowitz pushed his glass toward her and she decanted the remainder of the bottle into it. I flagged the waiter.

"The Cusan used an old analogy to explain what he was talking about. He said that learned ignorance is like knowledge of the Sun that is possessed by one who can see, as opposed to a blind person's. Let us say that a blind person asks his sighted friend about the brightness of the Sun. His friend investigates and reports that the Sun is so bright that it is impossible to look directly into it. The blind person then thinks that he has some knowledge of the brightness of the Sun, but he in fact remains completely ignorant. Now think of the sighted person. In order to answer the question, he has had to try to look into the Sun, and in doing that he has discovered that he cannot. He now has knowledge of his ignorance. He is certain that the Sun's brightness exceeds the limits of his vision."

"So you're saying Galileo is the blind man?"

"I am saying that Galileo and those in the Church who insisted on his prosecution were both blind."

I had to object. "But surely not Bellarmine. And probably not Urban either. They both very clearly understood the concept of learned ignorance."

"Yes, I agree with you. But it was clearly lower-ranking clerics who caused all the mischief, particularly in the first confrontation, the one over Copernicus. Bellarmine said very clearly that in the face of possible scientific proofs that the Earth revolved around the Sun, those passages in Scripture that seem to say the opposite should be interpreted with great circumspection, and—I think I can recall his exact words—'we should say that we do not understand, rather than affirm that what has been demonstrated is false.' And he was merely paraphrasing St. Augustine, who a thousand years earlier had said that if it happens that the authority of Sacred Scripture is set in opposition to clear and certain reasoning, this must mean that the person who interprets Scripture does not understand it correctly. It is not the meaning of Scripture that is opposed to the truth but the meaning that the interpreter has tried to give to it."*

---

*Here Sister Celeste has quoted Augustine verbatim, perhaps without realizing it. (The passage is from *Epistula* 143, n7 PL 33, col. 588.)

"But," I reminded her, "it was Urban himself who sent Galileo to trial, and who had him held under house arrest for the rest of his life. How do you account for that?"

"Right," echoed Berkowitz. "How do you account for that?"

"I would account for it by saying instead that it was Urban who saved Galileo from the full rigors of an Inquisition trial in the face of pressure from below and who mitigated his sentence when he had been convicted. Remember Bruno—he could have been burned. And I would further say that Urban did this in the light of his understanding of learned ignorance."

"Hmmm," I said.

Then Berkowitz sat up straight and reached into his jacket pocket, withdrawing three or four pages of text printed on computer paper. He had the look of a man who is about to lay down a straight flush.

"This afternoon," he announced, "I spent some time on the Internet. This is the text of John Paul II's speech on the occasion of the submission of the report of the Galileo Commission in 1992.[2] The title is 'Faith Can Never Conflict with Reason.' You're no doubt familiar with this, Sister, but I don't think that our friend here is. Allow me to read from it."

As he moved his plate and placed the papers on the table, flattening the folds with the side of his hand, I could see that he had highlighted several passages with a bright yellow marker.

"Proceed," Sister Celeste said, folding her arms.

"Part five," Berkowitz said. "'Thus the new science, with its methods and the freedom of research which they implied, obliged theologians to examine their own criteria of scriptural interpretation. Most of them did not know how to do so. Paradoxically, Galileo, a sincere believer, showed himself to be more perceptive in this regard than the theologians who opposed him.'

"Part nine: 'The majority of theologians did not recognize the formal distinction between Sacred Scripture and its interpretation, and this led them unduly to transpose into the realm of the doctrine of the faith a question which in fact pertained to scientific investigation.'

"Part twelve: 'The error of the theologians of the time, when they maintained the centrality of the Earth, was to think that our understanding of the physical world's structure was, in some way, imposed by the literal sense of Sacred Scripture. Let us recall the celebrated saying attributed to Baronius, *Spiritui Sancto mentem fuisse nos docere quomodo ad coelum eatur, non quomodo coelum gradiatur.** In fact, the Bible does not concern itself with the details of the physical world, the understanding of which is the competence of human experience and reasoning.'

"Now, with all due respect, Sister, I would call that a complete retraction and admission that Galileo was right." Berkowitz refolded the papers, his face a picture of smugness, an eyebrow raised, a faint, condescending smile. He was about to put them back in his pocket when Sister Celeste reached out to touch his arm.

"A moment, please. Continue reading from where you left us, if you don't mind."

Berkowitz hesitated only a moment before doing what he was told.

"'The error of the theologians . . .' blah blah blah, here we are: '. . . which is the competence of human experience and reasoning. There exist two realms of knowledge, one which has its source in Revelation and one which reason can discover by its own power. To the latter belong especially the experimental sciences and philosophy. The distinction between the two realms of knowledge ought not to be understood as opposition. The two realms are not altogether foreign to each other; they have points of contact. The methodologies proper to each make it possible to bring out different aspects of reality.'"

"Precisely," Sister Celeste said, and she reached for the papers. "May I?"

Berkowitz surrendered them to her reluctantly.

She quickly found the passage she was looking for. "'Like most of his adversaries'—I am reading from section five here—'Galileo made no distinction between the scientific approach to natural phenomena, and a reflection on nature of the philosophical order which that approach generally calls for. That is why he rejected the suggestion made to him to present the Copernican system as a

* "The purpose of Scripture is not to tell us how the heavens go, but how to go to heaven."

hypothesis, inasmuch as it had not been confirmed by irrefutable proof. Such, therefore, was an exigency of the experimental method of which he was the inspired founder.'"

"I read that, and I have no idea what it means," Berkowitz said.

"It means that Galileo gave no thought—no 'reflection'—to the philosophical assumptions and implications underlying his scientific approach. The assumption, for instance, of an ordered universe. In other words, he was not prepared to acknowledge the philosophical boundaries to scientific 'truth.' He refused to accept the argument presented by Bellarmine and Urban VIII and others that there is an underlying reality of which science can only provide a model, or a map. He thought he knew better. And what he thought he knew, I might add, was, as Rousseau says, dangerous."

Again, I had to object. "There, Sister, I think you, and perhaps even the pope, may be wrong."

Berkowitz guffawed. Sister Celeste, who had been sipping from her water glass, carefully placed it on the table and, smiling bemusedly, said, "A bold statement, sir! Explain yourself."

I poured myself some wine, collecting my thoughts.

"First of all, I believe that Galileo eventually did become aware of the implicit philosophical assumptions he'd been making. And second, I also believe that the pope is being too hard on the theologians of the time when he says most of them didn't know how to go about accommodating Scripture to new scientific fact. Bellarmine and Urban most certainly did, as you've agreed, and so did many others. The point is, there was no scientific 'fact' to be accommodated! Galileo had no proof that the Earth moved! He himself knew that, which is why he chose to write his book on the two world systems in the form of a dialogue rather than a straight scientific paper like Copernicus and Kepler had done. Or as he'd done himself with *The Starry Messenger* and all his mechanical papers. He needed to be able to use his magnificent rhetorical skills to make his case, because he just did not have the empirical evidence."

"That seems clear, I agree," said Sister Celeste. "But why do you say Galileo was aware of the philosophical problems with his views?"

"Because, for one thing, we know they were explained to him repeatedly by Bellarmine and Urban, and probably others, and he was not a stupid man. He must have understood, at least in principle. But more important, it seems to me that the whole problem of his abjuration makes no sense unless we assume he was telling the truth when he recanted, in other words that his abjuration was sincere."

Berkowitz was making protesting noises, but I held him off with a raised hand while I continued.

"We all know that Galileo recanted and read a statement of retraction. And we all know the myth about his saying under his breath 'Still it moves' is just that—a myth. He recanted, period. Now suppose we accept the idea that he did so insincerely, believing he actually had done nothing wrong. What does that make him? At best a hypocrite, at worst a perjurer. Probably both.

"The mythology excuses him of his perjury by saying he was terrified into doing what he did by fear of the rack. He was acting under duress. We all know, however, that there was no basis for any such fear. The so-called threat of torture was simply a legal formality in the Inquisitional process by that time, an antiquated and disreputable throwback to the idea of trial by ordeal. Even Bruno was never tortured. For another thing, Galileo had the protection of the Medicis in Florence. For another, Urban VIII was not about to allow him to be tortured, for both personal and political reasons. There were legal reasons as well. To torture so old and infirm a man would have been against the Inquisition's own rules.

"Torture was simply not in the cards, and Galileo must have known that. But let's give the myth the benefit of the doubt and assume he still harbored a niggling fear of the rack and that is what led him to recant. So what does *that* make him? A coward! And an irrational one at that. The myth attributes that to his advanced age. But are old men more cowardly than young?

"I don't know about you, but none of this fits my picture of Galileo. Brecht tries to explain it in his play by making Galileo a deeply flawed person, a sybarite addicted to the pleasures of the table and the cellar, an overly sensual man whose vivid imagination causes him to crumble morally at the thought of physical pain. But this is made entirely of whole cloth, for dramatic purposes. There is nothing in the historical record to support its premises—neither the real prospect of pain nor Galileo's physical cowardice.

Or, for that matter, his unnatural sybaritic tendencies. And there is certainly no suggestion of flawed character in his portraits—in fact, quite the contrary. There you see simply an intelligent, determined man. You certainly do not see a stupid, perjuring coward."

"Okay," Berkowitz said. "So what's the answer. How do you explain the abjuration."

"You're not paying attention," I said. "The explanation, I repeat, is that he was sincere. He meant what he said. He was making a sincere admission of error. And the reason he was able to do so was that he had been finally awakened to the idea of learned ignorance. He had certainly been exposed to the idea on many occasions, by both Bellarmine and Urban. The penny, or rather the scudo, finally dropped when he was brought up short by the Inquisition and forced to consider what he had done in writing the *Dialogue* in the way he did.

"If Galileo did have a serious character flaw, and again, I think we can all agree on this much, it was his oversize ego.* His lack of humility and charity got him into hot water all through his life, right? Well, I think his ego ran away with him in writing *Dialogue* the way he did. He was the blind man in the Cusan's story, as you said, Sister.

"If you read through the transcripts of the trial proceedings, you will find a letter Galileo wrote to the Cardinal Inquisitors admitting to exactly that. I think his words were that his mistake was one of 'vainglorious ambition.'

"Some historians have said this is simply evidence that a plea bargain was in the works and that he had been told he could weasel out of the charges with some such admission. But that seems to me too complicated an explanation for what is a very straightforward statement on his part. It makes more sense to me to take it at face

---

*A revelatory passage occurs in a letter Galileo wrote late in life, soon after he had become completely blind: "You may imagine the affliction this causes me when you stop to consider that the sky, the world, and the universe which, by my remarkable observations and clear demonstrations I had opened a hundred or a thousand times wider than anything seen by the learned of all the past centuries, is now diminished and restricted for me to a space no greater than that occupied by my own body."

value. To do otherwise is to accept him as a liar, on top of everything else.

"It seems to me that if you do not believe he was a perjurer or a hypocrite or a coward, or stupid or fatally narrow-minded, then the explanation I've just suggested is the only one that makes sense. He *had* to have been sincere. And I can see no other explanation for his sincerity than he genuinely accepted the flaw in his system and his own 'learned ignorance.' Certainly, there is no shame in submitting to superior arguments, which is what I believe he did."

Berkowitz was not convinced. "A very tidy argument," he said, but it was not a compliment. "In order to disagree with you, one has to accuse Galileo of being a coward and a liar. Very nice indeed. However, I seem to remember that Galileo was very specific in his statement of abjuration. Didn't he actually say that he no longer believed that the Earth moved and the Sun stood still? Whereas, of course, he certainly did still believe it. How could he possibly have been sincere in saying that?"

Sister Celeste, who seemed to have been lost in thought, suddenly brightened. "I think I may be able to help you with that, Mr. Berkowitz. What Galileo actually said in his statement—which I urge you to remember he himself had a hand in editing*—was that although he had been instructed by the Holy Office (in 1616) that he must abandon the Copernican hypothesis and not hold, defend, or teach it, verbally or in writing, because it was contrary to Holy Scripture, he printed a book in which he did exactly that. The book, he said, presented very strong arguments in favor of the Earth's movement and did not present the contrary arguments in a favorable light. That is to say, the book 'held,' or as we might say, 'upheld,' the Copernican theory. For that error, he said, he had

---

*This may or may not be true. The only evidence is a journal notation written at the time by G. F. Buonamici, whose source may have been Galileo himself: "[Galileo] begged the cardinals that if they insisted on proceeding against him in such a manner, they should at least leave out two points. . . . The first one was that he should not be made to say that he was not a good Catholic, for he was and intended to remain one despite all his enemies could say; the other that he would not say that he had ever deceived anybody, especially in the publishing of his book. . . ." However, the diary has proved unreliable in many details that may be checked against other documentary evidence.

been justly accused of vehement suspicion of heresy. He then went on to say—and I think I can remember his exact words here—he went on to say that he 'abjured, cursed, and detested' the 'aforementioned errors.' And that's as specific as he got.*

"Well, Mr. Berkowitz, as I understand that statement, the errors he refers to are those involved in disobeying the injunction of 1616. He does not say that he has suddenly reverted to the belief in the Ptolemaic system. He is saying that he sincerely regrets having published his *Dialogue* as it was written, that is, without having given proper weight to the arguments favoring Ptolemy in astronomy and especially of Bellarmine and Urban VIII on philosophy."

She paused for a sip of water, carefully eyeing Berkowitz, who had slouched low in his chair, his arms crossed.

"I can see I have not yet convinced you. Very well, there is one final argument I can make, though it is somewhat more obscure. You must understand first of all that in Galileo's time there was a well-known formula for distinguishing between knowledge derived from faith and knowledge derived from reason. It comes from St. Thomas Aquinas. It was understood that one could not hold the same truth by faith and reason at the same time. If you know something by reason, you cannot assent to it by faith. On the other hand, if you believe something—if you assent to it by faith—you do so because reason is incapable of telling you whether it is true.

"Now, this will be a stretch for you to understand, I fear. But it is possible for a true son of the Church, as Galileo was, to accept the beliefs prescribed by the Church as his own. A man like Galileo, to be sure, would accept such beliefs provisionally in cases where it seemed possible to confirm them by reason and science. But he would accept them, nevertheless. He would not reject them merely because they might at some time in the future be susceptible to scientific proof or disproof. To do so would make him a skeptic and not a Christian.

"Copernicanism, it seems to me, was such a case. The Church insisted it was incorrect. Galileo had a strong suspicion that it was correct. But he was unable, despite years of effort, to confirm his suspicions. In other words, he could not hold Copernicanism by

*The text appears on pages 256–57.

dint of reason, because he was unable conclusively to demonstrate it, and there was weighty evidence to oppose it. In 1633, the Inquisition forced him to acknowledge that fact and face up to it. And it asked him—demanded of him—to accept on faith the position of the Church. As a good and faithful son of the Church he was able to do so, because he understood, as Aquinas did, that what cannot be assented to by force of reason may still be accepted as true by an act of faith. In this case, acceptance was made easier by the weight of the authority of the Church. And so his abjuration was sincere."[3]

There was a long silence before Berkowitz spoke. The candles on the table were guttering in their pewter holder.

"I think that I'll surprise you both," he began, "by saying that the final argument is the one that makes most sense to me. I personally have trouble understanding how anyone could believe to be untrue something he was convinced *was* true, even though he couldn't prove it. But I accept, because you tell me it's so, that a good Christian is capable of doing this."

"Well," I offered, "it certainly has the virtue of being the simplest explanation. I also like it for theoretical reasons, that's to say because it fits with the Principle of Charity in interpreting history —it accepts his assertions at face value, at least as a starting point. And I think I can accept it as a historical fact that Galileo and his contemporaries were so deeply imbued with religious sensibility that it wouldn't have seemed an impossible feat to them."

"And there I can certainly agree with you," Sister Celeste said. "My goodness, it seems we are all in agreement!" She laughed. "Perhaps we should adjourn while we're ahead! I'm not sure I have the strength for further argument tonight."

"But we haven't ordered dessert," Berkowitz pointed out.

"Well, I think I do have strength for that, certainly," Sister Celeste said, and I signaled to our waiter, who, since we were now the only party remaining on the terrace, was enjoying a cigarette at one of the empty tables.

TWENTY-THREE

The Final Journey to Rome • The Inquisition
The Denial • The Confession • The Sentence

His delaying tactics exhausted, Galileo left for Rome on January 20, 1633, and arrived there on February 13, having spent some time in quarantine at the Tuscan border, a precaution made necessary by continuing outbreaks of the plague. He was not placed under arrest on his arrival, but was permitted to move into the Villa Medici as a guest of the Florentine ambassador. He was told to avoid socializing until the Inquisition called him for interrogation, but several of the cardinals of the Holy Office paid him informal visits, and "gentle conversation" was enjoyed at the table of the ambassador, Francesco Niccolini, and his wife, Caterina. Niccolini reported to Florence that "I think we have cheered up the old gentleman by showing him all that is being done for his cause; yet at times he comes back to finding this persecution very strange. I told him to show a will to obey and go very softly."[1]

Before many days had passed, the ambassador learned that the case against Galileo would hinge on one central accusation: that he had been ordered in 1616 not to discuss the question of the Earth's movement, either verbally or in writing, as the unsigned minute said. With this, the mood in the Villa Medici brightened considerably. Galileo insisted that those were not the terms of the injunction, and he had proof.

Niccolini began to press for Galileo to be allowed to return home, but on March 13 the pope informed him that the trial was to be held soon. A snippet of conversation from that audience, as preserved in Niccolini's diplomatic correspondence with Florence, is revealing:

> He [the pope] added that Signor Galileo had been his friend, that often they had dined familiarly at the same table, and that he was sorry to subject him to these annoyances, but that it was a matter of faith and religion. I think I remarked that when he was heard he would be able, without difficulty, to give all explanations requested. He answered that he will be examined in due time but that there is an argument to which they have never yet given an answer, and that is that God is all-powerful, and, if He is, why should we try to necessitate Him? I said that I did not know how to speak on those matters, but that I thought I had heard Galileo saying that he was willing not to believe in the motion of the Earth, but that as God could make the world in a thousand ways, so it could not be denied that He could have made it in this way too. He grew angry and replied that we should not impose necessity upon the Lord Almighty; and, as I saw him working himself up to a fury, I avoided saying more that might have hurt Galileo.

Once again Urban had made it abundantly clear that what was at issue in his confrontation with the scientist was not the problem of planetary motions but the much deeper question of the nature of truth and reality. By "imposing necessity" on God, he meant Galileo's insistence that there is a single and unique explanation to natural phenomena that may be understood though observation and reason and that makes all other explanations wrong.

At the same time Urban expressed his simmering anger with his former secretary Monsignor Ciampoli, on whose authority the book had been printed, linking him with Galileo in a way that suggests his suspicion of a conspiracy to betray him in allowing the *Dialogue* to be printed in the first place. "May God forgive Galileo for having intruded into these matters concerning new doctrines and Holy Scripture," he railed, ". . . and may God help Ciampoli too concerning these new notions, because he has a leaning toward them and is inclined to new philosophies."[2]

The processes of the Inquisition in the seventeenth century were highly formalized, and they merit some explanation. The word *trial* can be a misleading one when used in connection with its proceedings. A trial implies, to modern minds, an adversarial process in which the accused is accorded certain rights, including the right to be presumed innocent until proven guilty. There are other rights as well—for instance, the right to counsel; the right to know the nature of the crime you are accused of; the name of your accuser; the nature of the evidence against you and so on. None of this was true of the Inquisition as Galileo experienced it.

Prior to the revolution in political thought that followed the Scientific Revolution in the late seventeenth and eighteenth centuries, the idea of universal human rights was all but unknown, not simply in practice but in theory as well. Human rights, such as existed, were granted by superior authority: thus a king granted rights to his nobles, who in turn granted somewhat less generous rights to his vassals, and so on down the line to the lowly tenant farmer, who had very few secular rights indeed. In general, the primary concern was not with rights of the individual but with the maintenance of an orderly society.

One of the great moral accomplishments of the Christian Church of the Middle Ages was to ameliorate this rather harsh secular view of rights with the idea of equality among people of all stations. St. Thomas Aquinas spoke of the "dignity of man," suggesting essential equality among the virtuous.[3] This was not, however, an equality invested automatically and inalienably in every human being. Equal rights were accorded to men and women by the grace of God, and since they were conferred by the Deity, He could also take them away. Furthermore, this gift was accessible only to those baptized into the Church.* It followed that anyone who chose to leave the Church or to hold heretical views forfeited

---

*It is an idea that has an intellectual affinity with the concept of citizenship in the Roman Empire. A Roman citizen was defined as one who could claim kinship with the three original tribes of Rome, or an alien who had been granted citizenship. Citizenship was highly prized because it conferred partial immunity from legal torture or duress and the right to appeal decisions of any official in the empire to the Assembly in Rome.

any right to the protection that his privileges as a Christian might previously have afforded him. And this principle carried over in the area of what we would refer to as freedom of expression. This concept in its modern guise implies freedom to hold as valid and true, opinions contrary to those of the Church (or any other institution). But in Galileo's time, the Church maintained that to hold such views made one a heretic, or a non-Christian—and therefore one who had forfeited the rights bestowed at baptism. According to St. Thomas, the heretic was subject not just to excommunication but could lawfully be executed by properly constituted secular authority.

The trial process was by Galileo's time rigidly codified, books on procedure running to five hundred pages. Its object was less to ascertain specific offenses than to establish tendencies. Its ultimate goal was not punishment but the reclamation of lost souls. The Inquisitors hoped to effect a spiritual "cure." Confession was therefore deemed essential to a successful process, for both judicial and psychological reasons. Confession was also given priority status by the fact that the crimes being examined were intellectual and in many cases could not be satisfactorily demonstrated through material evidence.

Every person formally accused by the Inquisition was presumed to be guilty, and the judges were also the prosecutors. A written list of the accusations was given to the prisoner. This was not an adversarial process, since anyone defending the accused would be open to similar charges of heresy. However, the Inquisitors sometimes appointed an adviser to the accused from among their number.

The trials took place in secret, because the Church had no wish to further disseminate the heresies being examined. The accused was not told the names of his accusers so that they would not be placed at risk of revenge attacks by acting in the interests of the Church. Instead, the accused was given the opportunity to name his personal enemies and detail what each held against him. Thus, an accuser who was also named as an enemy might himself come under suspicion, and penalties for false accusations were potentially as severe as those facing the heretic.

The interrogation of the accused could be a protracted affair, and often took on the attributes of psychoanalysis.[4] Verbatim records were kept of each interview, and subsequent sessions would often refer to these transcripts. Frequently a complete life history would emerge.

Eventually the accused was formally charged with heresy, or "suspicion of heresy." The word *suspicion* was superfluous. While a heretic was one who had an unshakable commitment to one or more propositions that were in opposition to the doctrines of the Roman Catholic Church, suspects of heresy were those who occasionally stated such views, or who "keep, write, read, or give others to read books forbidden in the Index and in other particular Decrees . . ." There were degrees of suspicion of heresy—slight, violent, and vehement—depending on the seriousness of the offense. A plea of not guilty was not acceptable, and an accused who did not freely confess was returned to prison to reconsider.

Ultimately, the prisoner could be liable to torture, but only after all other avenues of inquiry had failed to resolve the case. The use of torture, too, was well codified. There were formally five degrees, ranging from being admonished, to being shown the implements, to being stripped and bound in the chamber, to being tied to the rack during questioning, to being actually subjected to pain. In theory, an accused could only be subjected to torture once, but this regulation was often breached through the expedient of calling each new session a "continuation" of the last.*

The penalties for those who confessed, or who had been convicted on other evidence, were solemnly pronounced on a

---

*The use of torture by the Inquisition was authorized for the first time by Innocent IV, in 1252. It was employed as an aid to the process of obtaining evidence, a confession being seen as the purest and most reliable evidence of all. Strict conditions were laid down to govern both circumstances of its use and its severity. For instance, it was not to cause permanent bodily damage or loss of life, and it was to be used only in grave cases where the presumption of guilt was very strong and all other means of inquiry had been exhausted. Its adoption by the Church coincided with the revival of Roman law in secular government throughout thirteenth-century Europe, in what is known as the "humanist revival." It appears in the civil code of Verona, for instance, in 1228 and in the "Sicilian" constitution of Holy Roman Emperor Frederick II in 1231. Its use by the Inquisition had all but disappeared, even in Spain, by the late eighteenth century, and it was formally banned by the Vatican in 1819.

Sunday in a church or other public place. These ranged in severity from penances, fasting, prayers, and pilgrimages to public scourgings and the compulsory wearing of a yellow felt cross sewn onto the clothing. Beyond these minor "penalties of humiliation" there were more severe punishments. The Inquisition could order imprisonment for any period up to and including life. And there were various degrees of imprisonment, the mildest of which amounted to a comfortable house arrest, and the most severe of which, the *murus strictus*, consisted of being placed in the deepest dungeon in single or double fetters and fed only bread and water.

The penalties having been read, the accused were called on one last time to confess and to abjure their crimes. If they complied, they were pronounced reconciled with the Church and sent to prison at the pleasure of the Inquisition. If they refused, they were handed over to civil authorities to be burned, and in burning, purified.[5]

The first of three interrogations of Galileo by the Inquisition took place on April 12, 1633. As custom prescribed that the accused be strictly secluded during trial proceedings, he was taken from the Villa Medici to the palace of the Holy Office near the Vatican, where he was given a comfortable apartment. The hearing was conducted by the commissary general of the Inquisition, Father Vincenzo Maculano da Firenzuola and his staff of two, in the commissary's apartment in the palace. Galileo was asked about the events of 1616, and he admitted that he had been given a warning by Cardinal Bellarmine not to hold or defend the Copernican theory "taken absolutely," although he was free to discuss it as a hypothesis, that is, as a mathematical model designed to account for observed appearances. He then, to his questioner's apparent surprise, produced a copy of the letter he had obtained from Cardinal Bellarmine as verification. (Bellarmine himself had died in 1621.) This was entered as evidence, and the commissary went on to question him about whether he had been given any additional instructions by the Dominican commissary present at the time of his meeting with Bellarmine, in particular those detailed in the unsigned minute that enjoined him from holding, teaching, or defending the thesis that the Sun is stationary and the Earth moves "in any way, by word or in writing." Galileo denied any memory of such a special injunction.

Galileo was then asked why, in view of his having been warned against holding or defending the Copernican thesis, he had not asked permission to write his book, and why he had not mentioned the injunction when he sought permission to have the book printed. Galileo replied that he had been secure in the knowledge that the book in no way violated Bellarmine's warning, and that therefore there had been no need to seek permission or notify the censors: "I did not happen to discuss that command with the Master of the Sacred palace when I asked for the imprimatur, for I did not think it necessary to say anything, because I had no doubts about it; for I have neither maintained nor defended in that book the opinion that the Earth moves and that the Sun is stationary but have rather demonstrated the opposite of the Copernican opinion and shown that the arguments of Copernicus are weak and not conclusive."[6]

This last statement must have amazed the Inquisitors. The accused in these proceedings was not expected to challenge the prosecution, and Galileo was flouting the stern advice he had been given by Ambassador Niccolini to tread softly and try to give his questioners the answers they were looking for. The purpose of the trial, Niccolini would have reminded the scientist, was not to establish guilt—that had already been decided. The *Dialogue* was there for everyone to read, and it had been examined by a panel who had concluded unanimously that it both held and defended Copernicanism. The purpose of the trial was to bring the accused to recognize his transgression and reconcile himself with the Church.

Had Galileo freely admitted at this stage that he had overstepped the bounds of the injunction given him by Bellarmine, it is likely the outcome of the trial would have been different. He might, for example, have been given the opportunity to rewrite offending passages of the book to make it clear he was treating Copernicanism as a mathematical hypothesis. As it was, the Inquisition was now faced with the ticklish question of how to extract the necessary confession of guilt from so elderly and illustrious a prisoner.

In the face of Galileo's denial, the proceedings were adjourned while a second panel of Counselors to the Holy Office was asked to read the book and give their opinions whether it held and defended

Copernicanism. Two of the three had served on the initial commission of inquiry. Their reports were ready within a week, and once again the verdict was unanimous and went against Galileo. The book, they said, violated even the milder interpretation of the 1616 injunction contained in Cardinal Bellarmine's letter, and furthermore grounds existed for "vehement suspicion" of heresy on the part of the author. Each provided long lists of quotations to support his position. The Jesuit reviewer, Melchior Inchofer, seemed to take personal offense at Galileo's rhetorical style. If the author had not held the Copernican position, he pointed out, he would not have written the "Letter to the Grand Duchess Christina," "nor would he have held up to ridicule those who maintain the accepted opinion, and as if they were dumb mooncalves or described them as hardly deserving to be called human beings.* Indeed, if Galileo had attacked some individual thinker for his inadequate arguments in favor of the stability of the Earth, we might still put a favorable construction on his text, but as he declares war on everybody and regards as mental dwarfs all who are not Pythagorean or Copernican, it is clear enough what he has in mind."†[7]

With this additional testimony in hand, the commissary general asked the Inquisition panel's ten cardinals' permission to "treat extrajudicially with Galileo, in order to render him sensible of his error and bring him, if he recognizes it, to a confession of same." In other words, he wanted to try to explain to the old man his perilous position vis-à-vis the Inquisition, and to talk some sense into him. The permission granted, he met with Galileo the next day, and was able to report a hopeful result in a note to Cardinal Francesco Barberini, the pope's nephew and personal liaison with the Inquisition panel: "I entered into discourse with Galileo yesterday afternoon, and after many and many arguments and rejoinders had passed between us, by God's grace, I attained my object, for I brought him to a full sense of his error, so that he clearly recognized that he had erred and had gone too far in his

* The offending passage from the *Dialogue* is "I have met such arguments that I blush to rehearse them, not so much to spare the shame to their authors, the names of whom might remain perpetually concealed, as because I am ashamed so deeply to degrade the honor of mankind."

† The Pythagoras reference can be seen as evidence that Inchofer, and presumably the other consultants, understood the real significance of the offense.

book. And to all this he gave expression in words of much feeling, like one who experienced great consolation in the recognition of his error, and he was also willing to confess it judicially. He requested, however, a little time in order to consider the form in which he might most fittingly make the confession. . . ."[8] It may be that the commissary general's personal background was instrumental in his success. A Dominican priest, he was also a military engineer who was sympathetic to the Copernican view himself. In any case it appeared that the case could now be satisfactorily concluded.

> I trust that His Holiness and your Eminence will be satisfied that in this way the affair is being brought to such a point that it may soon be settled without difficulty. The court will maintain reputation; it will be possible to deal leniently with the culprit; and whatever the decision arrived at, he will recognize the favor shown him, with all the other consequences of satisfaction herein desired. Today I think of examining him in order to [officially] obtain the said confession; and having, as I hope, received it, it will only remain to me further to question him with regard to his intention and to receive his defense plea; that done, he might have [his] house assigned to him as a prison, as hinted to me by your Eminence, to whom I offer my most humble reverence.[9]

The case would be resolved in exactly this manner.*

On April 30, Galileo was interviewed once again in the presence

---

*It has been suggested by de Santillana and other historians that the commissary's initiative was intended to be part of a "plea bargain" settlement, under which Galileo would confess to having unintentionally overstepped the prohibition, in return for a lenient sentence, or even a dropping of the case by the Holy Office. This is the purest speculation, and it seems highly improbable, if only because Galileo's guilt was not in question at the trial—it had been established beforehand in the accepted manner of Inquisition proceedings. Nor could it be seriously questioned on the basis of the evidence of the book itself. And Urban, as he made clear to Niccolini, felt he had no choice but to impose a significant penalty. It makes much more sense to take events at their face value: Galileo was warned how serious his position was and that the only way to avoid the severe sanctions imposed on all those who refused to confess and reconcile themselves with the Church was to make a clear admission of guilt. He accepted the warning.

of the commissary general and his staff. He made a statement, which was recorded as follows:

> In the course of some days' continuous and attentive reflection on the interrogations put to me on the twelfth of the present month, and in particular as to whether, sixteen years ago, an injunction was intimated to me by order of the Holy Office, forbidding me to hold, defend, or teach 'in any manner' the opinion that had just been condemned—of the motion of the Earth and the stability of the Sun—it occurred to me to reperuse my printed *Dialogue*, which for three years I had not seen, in order carefully to note whether, contrary to my most sincere intention, there had, by inadvertence, fallen from my pen anything from which a reader or the authorities might infer not only some taint of disobedience on my part but also other particulars which might induce the belief that I had contravened the orders of the Holy Church. . . . And as, owing to my not having seen it for so long, it presented itself to me, as it were, like a new writing and by another author, I freely confess that in several places it seemed to me set forth in such a form that a reader ignorant of my real purpose might have had reason to suppose that the arguments brought on the false side, and which it was my intention to confute, were so expressed as to be calculated rather to compel conviction by their cogency than to be easy of solution. Two arguments there are in particular—the one taken from the solar spots, the other from the ebb and flow of the tide—which in truth come to the ear of the reader with far greater show of force and power than ought to have been imparted to them by one who regarded them as inconclusive and who intended to refute them, as indeed I truly and sincerely held and do hold them to be inconclusive and admitting of refutation. And, as an excuse to myself for having fallen into an error so foreign to my intention, not contenting myself entirely with saying that, when a man recites the arguments of the opposite side with the object of refuting them, he should, especially in writing in the form of dialogue, state these in their strictest form and should not cloak them to disadvantage of his opponent—not contenting myself, I say, with this

> excuse, I resorted to that of the natural complacency which every man feels with regard to his own subtleties and in showing himself more skillful than the generality of men in devising, even in favor of false propositions, ingenious and plausible arguments. With all this, although with Cicero *'avidior sim gloriae quam sat est,'* if I had not to set forth the same reasonings, without doubt I should so weaken them that they should not be able to make an apparent show of that force of which they are really and essentially devoid. My error, then, has been—and I confess it—one of vainglorious ambition and of pure ignorance and inadvertence."

Galileo then offered to add one or two new chapters to his book to correct the problem: "I promise to resume the arguments already brought in favor of the said opinion, which is false and has been condemned, and to confute them in such most effectual manner as by the blessing of God may be supplied to me."

The offer was ignored, and the prisoner, by now much fatigued by stress, was allowed to return to the more congenial surroundings of the Villa Medici. Niccolini reported that Galileo was "more dead than alive" when he appeared at the villa that day.

The next interview was held on May 10, and at that time Galileo was informed that he would be allowed eight days to prepare and present his defense, should he wish to submit one. Galileo at once presented the prosecutor with a handwritten document that had doubtless been worked out with the assistance of Ambassador Niccolini. He again denied any recollection of the special injunction contained in the unsigned minute. Since the warning he did recall having been given by Cardinal Bellarmine was the same as that issued to all Catholics at the time of the 1616 finding against Copernicanism by the Holy Office, he had seen no need to mention it to the censor, who as a functionary of the Holy Office would surely have been aware of it himself. The "faults" found in the book were not "artfully introduced with any concealed or other than sincere intention but have only inadvertently fallen from my pen, owing to a vainglorious ambition and complacency in desiring to appear more subtle than the generality of popular writers, as indeed in another deposition I have confessed; which fault I shall be ready to correct with all possible industry whenever I may be commanded or permitted by Their Most Eminent Lordships." Finally, he asked for clemency on grounds of his advanced age and infirmities.

On June 16, the final disposition of the case was announced by the Holy Office. Galileo was to be interrogated one last time in order to determine his intention in writing the book. If he satisfied the Inquisitors, he was to appear before a plenary assembly of the Congregation of the Holy Office to abjure, and to be sentenced to imprisonment and ordered "not to treat further, in whatever manner, either in words or in writing, of the mobility of the Earth and the stability of the Sun. . . ."[10] His book was to be prohibited, not simply withdrawn for correction.

Whether this was a more severe sentence than Galileo had been led to expect by the commissary-general at the time of their interview is impossible to know. It does not seem to have surprised Ambassador Niccolini, who had predicted as much in a report to Florence a month earlier, after his regular audience with the pope.[11] And in a subsequent interview, Urban VIII told Niccolini that Galileo would in all likelihood be imprisoned for a time, for having transgressed the injunction of 1616. "However," the pope had added, "after the publication of the sentence We shall see you again, and We shall consult together so that he may suffer as little distress as possible, since it cannot be let pass without some demonstration against his person [i.e., punishment]." The pope went on to say that "he would at any rate be sent for a time to some monastery, like Santa Croce,* for instance; for he really did not know precisely what the Holy Congregation might decree, but it was going along unanimously . . . in the sense of imposing a penance."[12]

Galileo appeared before the commissary-general for the last time on June 21, to be questioned as to "intention"—that is, his true convictions. As was prescribed in Inquisition procedures, the interrogation was done under formal, verbal threat of torture. However, this was understood on both sides to be nothing more than a procedural formality, as the same regulations prohibited the torture of anyone of Galileo's age and state of health. The purpose was to ensure a truthful statement.

---

*The abbey of Santa Croce in the duchy of Urbino. The poet Dante is said to have lived there for some time.

Galileo was asked "whether he holds or has held, and for how long, that the Sun is the center of the world and the Earth is not the center of the world but moves also with diurnal motion." He replied that for a long time prior to the injunction of 1616 and the decision of the Congregation of the Index he had been undecided whether the Copernican or the Ptolemaic systems were "true in nature." But "after the said decision, assured by the prudence of the authorities, all my uncertainty stopped, and I held, as I still hold, as most true and indisputable, Ptolemy's opinion, namely the stability of the Earth and the motion of the Sun." As to his intentions in writing and publishing the *Dialogue*, "I did not do so because I held the Copernican doctrine to be true. Instead, deeming only to confer a common benefit, I set forth the physical and astronomical reasons that can be advanced for each side; I tried to show that neither set of arguments has the force of conclusive demonstration. . . ." The commissary chose not to press him on either point, although he certainly had the ammunition to do so in the reports of the three consultants and the preliminary commission of inquiry. Galileo concluded, "I do not hold the opinion of Copernicus, and I have not held it after being ordered by the injunction to abandon it. For the rest, I am here in your hands to do with me what you please."

He had passed his final test: in the face of the threat of torture, however notional, he had continued to deny any doctrinally false belief or malicious intent, and his statement was accepted.

The statement was, of course, manifestly untrue—he had been a confirmed Copernican throughout.* This presents obvious difficulties for those who would argue for Galileo's sincerity in the process. The problem is perhaps not insurmountable, given that Galileo made no such statements in his public confession and abjuration. It may be that he saw the necessity of doing so in his examination as to intent, as strategically unavoidable. He may well

---

*The best evidence of this is of course contained in his *Dialogue*. There is also correspondence, including the letter to Diodati quoted on page 224: "I have taken up work again on the *Dialogue of the Ebb and Flow of the Sea*, which was left aside for three years, and with God's grace I have found the right line, which ought to allow me to terminate it within the winter; it will provide, I trust, a most ample confirmation of the Copernican system." He is also explicit in his Copernicanism in his *Sunspot Letters* and the "Letter to the Grand Duchess Christina."

have been advised that this and no other response could be accepted by the Holy Office. The documents relating to the trial, including this one, were kept strictly secret, so Galileo knew there would be no risk of his statement reaching the public, as his more circumspect abjuration would.*

---

*See Appendix: On the Question of Galileo's "Perjury," p. 279.

TWENTY-FOUR

The Judgment and Abjuration
The Unspoken Issues

The judgment against Galileo was delivered on June 22, 1633, in a large, well-lit, marble-floored room in the splendid Renaissance convent of Santa Maria sopre Minerva, close to the ancient Pantheon in central Rome. The proceedings unfolded beneath a spectacular ceiling fresco that depicted the bloody victory of northern French Crusaders over the Albigensian heretics at Muret in 1213. The Albigensians, or Cathars, of present-day Languedoc and Provence had raised complaints against the Church that were remarkably similar to those voiced four hundred years later by Luther and Calvin in the Protestant Reformation. They demanded an end to the sale of indulgences, a less venal and more literate clergy, and the right to read and interpret the Bible without the mediation of Church fathers. Their priesthood, the Perfecti, was rigorously ascetic, and many of their most prominent supporters were women of the southern nobility. Deemed heretical because it denied several basic tenets of orthodox Christianity, including the divinity of the historical Christ, the movement was met ultimately with military force, but not before it had prompted institutional reforms, the clarification of dogma, and the creation of the mendicant preaching orders, the Franciscans and the Dominicans. It was in the wake of Muret and other military victories that the papacy created the Inquisition and put it in the hands of the Dominicans, the "black-and-white hounds of the Lord," as a means of ferreting out and prosecuting those Cathars who had survived the holocaust of the Crusade. It was a remarkably effective instrument, and achieved virtually complete success. There could scarcely have been a more appropriate decoration for a room in which an ultimately disastrous chapter in the militant orthodoxy of the Counter-Reformation was being written by Galileo and his Inquisitors, just as four hundred years earlier the Albigensian Crusade had provided a different, and bloodier, expression of militant reform.

Only seven of the ten cardinals of the Inquisition signed the judgment delivered in the Minerva that day. The absent signatures include that of Francesco Barberini, who is known to have been sympathetic to Galileo and would later intercede on his behalf with his uncle, the pope. On this evidence some historians have built a speculative case for a split in the ranks of the Inquisition panel, with those favoring a harsh judgment on the one side and those arguing for leniency on the other. However, while there may have been several cardinals who were sympathetic to either Galileo personally or to the Copernican hypothesis, or both, it is far from clear whether they would have objected to his treatment at the hands of the court. It was, in fact, a rather lenient judgment, given both the nature of the offense and Galileo's hedging admission of guilt in his final interrogation.* (Another of the missing signatures belonged to Cardinal Gaspar Borgia, who had earlier in the year accused the pope of lacking crusading zeal. Borgia may have abstained in solidarity with the grand duke of Tuscany, Galileo's patron and protector, who had close ties with Spain and the empire. The third was that of Cardinal Laudivio Zacchia, whose views are unknown.)

The judgment, as it was read out before the court assembled beneath the lurid scene of the Battle of Muret, reviewed the facts of the case in detail:

---

*It is perhaps worth reiterating here that the question of whether Galileo and others like him should have been free to say and publish whatever they wished is one that is relevant only in hindsight. In 1633 the notion of freedom of expression was as alien to common wisdom as the idea of the separation of Church and state.

> . . . and whereas a book appeared here recently, printed last year in Florence, the title of which shows that you were the author . . . and whereas the Holy Office was afterward informed that through the publication of the said book the false opinion of the motion of the Earth and the stability of the Sun was daily gaining ground, the said book was taken into careful consideration and in it there was discovered a patent violation of the aforesaid injunction that had been imposed upon you, for in this book you have defended the said opinion previously condemned and to your face declared to be so, although in the said book you strive by various devices to produce the impression that you leave it undecided, and in express terms as probable; which, however, is a most grievous error, as an opinion can in no wise be probable which has been declared and defined to be contrary to divine Scripture. . . .[1]

The letter from Cardinal Bellarmine that Galileo had entered in his defense was dismissed as merely increasing his culpability, since it stated clearly that the Copernican hypothesis had been found to be contrary to Scripture and "you have nevertheless dared to discuss and defend it and to argue its probability." Galileo's initial reluctance to admit guilt was noted: "And whereas it appeared to us that you had not stated the full truth with regard to your intention [in publishing], we thought it necessary to subject you to a rigorous examination at which . . . you answered like a good Catholic."

And then the sentence:

> We say, pronounce, sentence, and declare that you, the said Galileo, by reason of the matters adduced in trial, and by you confessed as above, have rendered yourself in judgment of this Holy Office vehemently suspected of heresy . . . and that consequently you have incurred all the measures and penalties imposed and promulgated in the sacred canons and other constitutions, general and particular, against such delinquents. From which we are content that you be absolved, provided that, first, with a sincere heart and unfeigned faith, you abjure, curse, and detest before us the aforesaid errors and heresies and every other error and heresy contrary to the Catholic and

Apostolic Roman Church in the form to be prescribed by us for you.

And in order that this your grave and pernicious error and transgression may not remain altogether unpunished and that you may be more cautious in the future and an example to others that they may abstain from similar delinquencies, we ordain that the book of the "Dialogue of Galileo Galilei" be prohibited by public edict.

We condemn you to formal imprisonment in this Holy Office during our pleasure, and by way of salutary penance we enjoin that for three years to come you repeat once a week the seven penitential Psalms. Reserving to ourselves liberty to moderate, commute, or take off, in whole or in part, the aforesaid penalties and penance.

Galileo knelt and read the statement of abjuration that had been prepared for him:

I, Galileo, son of the late Vincenzo Galilei, Florentine, aged seventy years . . . swear that I have always believed, do believe, and by God's help will in the future believe all that is held, preached, and taught by the Holy Catholic and Apostolic Church. But whereas—after an injunction had been judicially intimated to me by this Holy Office to the effect that I must altogether abandon the false opinion that the Sun is the center of the world and immovable and that the Earth is not the center of the world and moves and that I must not hold, defend, or teach in any way whatsoever, verbally or in writing, the said false doctrine, and after it had been notified to me that the said doctrine was contrary to Holy Scripture—I wrote and printed a book in which I discuss this new doctrine already condemned and adduce arguments of great cogency in its favor without presenting any solution of these, I have been pronounced by the Holy Office to be vehemently suspected of heresy, that is to say, of having held and believed that the Sun is the center of the world and immovable and that the Earth is not at the center and moves. Therefore, desiring to remove from the minds of your Eminences, and of all faithful Christians, this vehement

> suspicion justly conceived against me, with sincere heart and unfeigned faith I abjure, curse, and detest the aforesaid errors and heresies and generally every other error, heresy, and sect whatsoever contrary to the Holy Church, and I swear that in future I will never again say or assert, verbally or in writing, anything that might furnish occasion for a similar suspicion regarding me. . . . I, the said Galileo Galilei, have abjured, sworn, promised, and bound myself as above; and in witness of the truth thereof I have with my hand subscribed the present document of my abjuration and recited it word for word at Rome, in the convent of the Minerva, this twenty-second day of June, 1633.

And by this statement Galileo was, according to the four-hundred-year-old formula of the Inquisition, reconciled with the Church and taken back into the fold.

Historians have found irony in the fact that even though Copernicus himself had never been accused of heresy, and his book had been merely withdrawn temporarily for correction, Galileo stood "vehemently suspected of heresy" and saw his book banned—all for arguing the Copernican case. But this observation merely reinforces the point that the Church's deepest concerns were not focused on the Copernican system per se, but on the much broader issue of Galileo's mechanistic, materialist philosophy that insisted, in essence, that seeing is believing, whereas the Church held, with Plato and St. Augustine and subsequent generations of scholars, that there is much believing involved in seeing. On so seemingly petty a difference of opinion, one world was lost and another gained. If there is irony to be found in the trial, it is in the fact that in this climactic event of the historic confrontation, the real issues, the issues Urban VIII had called "perverse in the extreme," and "the most dangerous that can be stirred up," went unmentioned.*

Galileo's trial had focused instead on the narrow issue of papal authority. It was a subject very much on the front burner in Urban VIII's Rome, with the dangerous challenges of the Cardinal Borgia

---

* See Chapter Nineteen for the context of these quotations.

affair and the Urbino dispute still fresh in memory. The *Dialogue* had been shown to have directly contravened an injunction issued in the name of the pope. Not just Galileo, but Ciampoli and perhaps even Riccardi—Tuscans all, trusted countrymen—appeared to have conspired to publish the book in a form that made a mockery of the pope's intellectual authority and undermined his temporal power.

But there was a much larger issue of authority that, like the underlying philosophical issues, went unaddressed. The book's purpose had been clear in Galileo's mind from the time he first conceived of it, decades earlier. It was to force the Church to abandon its ancient attachment to the thought of Aristotle and accept the new mathematical realism and experimental science. The lever Galileo used was the threat of the exposure of Scripture to ridicule, unless it was promptly reinterpreted to fit the new scientific knowledge of the Earth's movement.

The problem was that as Galileo framed the dispute, its resolution demanded nothing less than the Church's acquiescence in the dismembering of philosophy into two separate disciplines, moral and natural. And it demanded religion's complete withdrawal from the field of science and the interpretation of scientific knowledge. Here was the ultimate challenge to the Church's authority, beside which all others paled. The trial of 1633 confronted only its shadows.

TWENTY-FIVE

The Aftermath • *Return to Arcetri*
Discourses on Two New Sciences
A Visit from Hobbes • *Milton's Tribute*

The terms of Galileo's "formal imprisonment" as prescribed in his sentence amounted to house arrest. He first returned to the Villa Medici where he spent about ten days sequestered in the palace and its gardens. After that, he was released to the custody of Anscanio Piccolomini, the archbishop of Siena, who provided an apartment in that city's magnificent episcopal palace for the old man.

Scion of a family that had produced two popes, Pius II and III, and a skilled mathematician in his own right, Piccolomini was a man of liberal bent, and an admirer of Galileo's. A salon atmosphere quickly developed around the old scientist, and within weeks he was at work on what would be his greatest book, *Discourses on Two New Sciences*. It was a decidedly more agreeable confinement than the rigorously disciplined monastery existence that had earlier been proposed—the pope had kept his word in exercising leniency.

After about five months in Siena, Galileo was permitted to return to his villa at Arcetri just outside Florence, and he was once again able to visit his daughters at their nearby convent of San Mattero. When his eldest daughter and cherished correspondent, Sister Maria Celeste, died at thirty-three soon after his return, in April 1634, he was inconsolable.

In 1636 Luis Elzevir, the Dutch publisher, visited Galileo at his villa to discuss publishing the old scientist's book on mechanics, so long in gestation. It was arranged that the manuscript would be delivered to him a chapter at a time, as it was completed, by a trusted friend of Galileo's in Venice. The first copies of the book appeared in 1638. Galileo's long dedication to the French ambassador to Florence is a transparent fabrication of the events leading to its publication:

> . . . notwithstanding the fact that I myself, as you know, being confused and dismayed by the ill fortune of my other works had resolved not to put before the public any more of my labors. Yet, in order that they might not be completely buried, I was persuaded to leave a manuscript copy in some place. . . . And thus having chosen, as the best and loftiest such place, to put this into your excellency's hands . . . I was . . . preparing some other copies [of the manuscript] to send to Germany, Flanders, England, Spain, and perhaps also to some place in Italy, when I was notified by the Elzevirs that they had these works of mine in press, and that I must therefore decide about the dedication and send them promptly my thoughts on the subject. From this unexpected and astonishing news, I concluded that it had been your excellency's wish to elevate and spread my name by sharing various of my writings, that accounted for their having come into the hands of those printers who, being engaged in the publication of other works of mine ["Letter to the Grand Duchess Christina"] wished to honor me by bringing these also to light at their handsome and elaborate press. . . .[1]

The disingenuousness of this dedication, like the fawning letter to the Grand Duke Cosimo de' Medici in which, as a newly minted scientific celebrity, he begged to be appointed chief mathematician and philosopher to the Florentine Court, make Galileo a difficult hero, though a fascinating human being.

The ailments of old age and advancing blindness did not prevent Galileo from continuing his scientific writings, dictating to his young secretary, student, and eventual hagiographer, Vincenzo

Viviani. And before he was completely blind, he made one last astronomical discovery—he noticed that the Moon wobbles slightly on its axis, the lunar libration. His old friend and former student Benedetto Castelli gained permission from the Holy Office to visit him to study the motions of the Jovian moons and to offer spiritual advice. His son Vincenzo visited with his new grandson, and father and son worked on plans for a pendulum clock.

The influential English philosopher Thomas Hobbes (1588–1679) made a pilgrimage to Arcetri in 1636 while on the grand tour with his student, the future earl of Devonshire. Hobbes, best known for his political philosophy, was not much interested in astronomy, but Galileo's mathematical materialism fascinated him, and his biographers attribute to Galileo a seminal influence on his thinking. Hobbes was also a close associate of René Descartes, another of the founders of modern philosophy. A thoroughgoing materialist and reductionist, Hobbes proclaims at the beginning of his most famous book, *Leviathan* (1651), that life is nothing but motion and matter. Reasoning is not innate but is developed through experience. It is a form of mathematics, which is the one true science. Our thought processes are not arbitrary but are governed by laws of association. Just as laws of the behavior of inanimate bodies are deduced from mathematical models of the relationships of space, time, force, and power, so should it be possible to understand the basic properties of human societies, since people are essentially machines. There is no room for free will in Hobbes's philosophy, and very little for God, who is assigned the task of Prime Mover. In this, Hobbes influenced another of the early modern philosophers, Baruch Spinoza. Spinoza (1632–77), in his metaphysics, used the terms *God* and *nature* interchangeably. His naturalized God had no desires or purposes, and so human ethics could not properly be derived from divine purpose or commandment.

In Hobbes, and in the materialist philosophy he helped to popularize, much that is implicit in Galileo is made explicit.* In particular, the impact on ethics, psychology, and politics is more clearly

*Reading of Hobbes was banned both by the Congregation of the Index in Rome and by Oxford University.

defined. At the same time the central flaw in Galileo's mathematical realism is transferred over into these areas. It was assumed that reason alone can establish with certainty the nature of existence, and therefore human actions and interactions can, in principle, be reduced to mathematical formulas. The fallacy of this, so well understood in religious terms by the Church of Bellarmine and Urban VIII, would later be established philosophically by David Hume (1711–76) and more definitively by Immanuel Kant (1724–1804). It is, briefly, this. Hobbes believed that the facts of human behavior as they may be observed can yield not only descriptive generalizations of the kind scientists like to make but also prescriptive generalizations or principles for correct behavior—rules of morality. This is a delusion. The transition from what is to what ought to be can never be made simply by observing and cataloging human behavior. What "ought to be" depends on some notion of human possibility, as distinct from actuality. People ought to be, in other words, what they are ideally capable of being. And so in order to have an idea of what constitutes moral behavior it is necessary to first conceive a priori what an ideal human might be.*

But by then scientism had the upper hand, and it was the antimaterialist philosophers who found themselves under siege at every turning from the new orthodoxy, and largely ignored by the educated elites to whom Galileo had so deliberately appealed.

In October 1641 Galileo's little household at Arcetri was joined by a brilliant student of Castelli's, Evangelista Torricelli, a self-proclaimed "Galileist." In a letter written the previous month, Galileo had invited him up from Rome: "I hope to enjoy your company for some few days before my life, now near an end, is finished . . . to discuss with you some relics of my thoughts on mathematics and physics and to have your aid in polishing them, that they may be left less messy to be seen with other things of mine."[2]

Galileo wrote to a friend: "I have in mind a great many miscellaneous problems and questions, partly quite new and partly dif-

*It is not enough to merely catalog the "best" of observed human behavior and draw a picture of an "ideal man" based on that. Such an ideal is nothing more than an observation, a sociological fact. There is no basis for stating that it is a prescription for how people ought to act in all places at all times, which is what morality is all about.

ferent from or contrary to those commonly received, of which I could make a book more curious than the others written by me; but my condition, besides blindness on top of other serious indispositions and a decrepit age of seventy-five years, will not permit me to occupy myself in study. I shall therefore remain silent, and so pass what remains to me of my laborious life, satisfying myself in the pleasure I feel from the discoveries of other pilgrim minds."[3] He died January 8, 1642, of fever and kidney disease. His students and his son were at his bedside.

One of Galileo's last visitors was the young English poet John Milton. Their meeting was fraught with mythological significance: Milton, destined to become second only to Shakespeare in the ranks of English poets, violently anti-papist, the great defender of human rights and republicanism whose political masterwork, *Areopagitica*, would be a brilliant argument for freedom of the press in the face of Puritan censorship; and the old scientist who had been muzzled by entrenched authority in Rome. In *Areopagitica*, Milton wrote of their encounter:

> . . . I could recount what I have seen and heard in other countries, where this kind of inquisition tyrannizes; when I have sat among their learned men, for that honor I had, and been counted happy to be born in such a place of philosophic freedom as they supposed England was, while they themselves did nothing but bemoan the servile condition into which learning amongst them was brought; that this was it which had dampened the glory of Italian wits; that nothing had been there written now these many years but flattery and fustian. There it was that I found and visited the famous Galileo, grown old, a prisoner to the Inquisition, for thinking in astronomy, otherwise than the Franciscan and Dominican licensers thought.[4]

On these much-quoted lines of Milton's political rhetoric is anchored the conventional wisdom that with Galileo's sentencing, original thought in Italy came to a halt. There are a number of reasons why this notion should be challenged. First, the charge of heresy was an ever-present risk to proponents of the new mathematical, mechanistic philosophy not only in Catholic Italy but in Protestant Europe as well, Holland being the single honorable exception to the rule. Even in liberal England, Hobbes himself was threatened with the burning of his books, not to mention his person. Few historians of the period would be willing to argue that Galileo might have fared better under a Calvinist or Lutheran regime.

Second, scientific inquiry did in fact continue in Italy after the Galileo affair. Galileo had not been long dead when there sprang up in Florence the Conversazione Filosofica, a society of the ablest literary and scientific minds of the time, which met regularly in the ducal palace. In 1657 the Academia del Cimento (Academy of Experiment), Europe's first academy of experimental science, was founded by Prince Leopold of Tuscany under the patronage of Grand Duke Ferdinand II, both former students of Galileo's. This was the prototype for the Royal Academy in London and the Académie Française in Paris. At its first meeting it adopted a constitution that bound it to "investigate nature by the pure light of experimental facts." In Rome, Galileo's beloved Lincean Academy functions to this day.

While it is an undeniable fact of history that the artistic and scientific brilliance with which Italy had shone all through the Renaissance and the Reformation dimmed sharply toward the middle of the seventeenth century, the decline cannot reasonably be attributed to the chilling effect of the Inquisition's action against Galileo. It is rather a function of the historic shift in economic and political power away from Italy, Spain, and the Holy Roman Empire in favor of the new nation-states of the northern Atlantic seaboard, the main beneficiaries of the migration of commerce from the Mediterranean to the Atlantic. Throughout Europe, the seventeenth century was distinguished by a long series of economic crises that came with the cooling-off of the previous century's fabulous global expansion. Toward the end of the century, further economic disruption was caused by the introduction of new industrial technologies. Italy's industrial centers were especially hard hit, and the country's best and brightest were drawn to economic opportunities outside their war-exhausted, economically depressed homeland.[5]

Certainly, intellectual freedom is a precious right, greatly to be desired in any era. The Church, for all its well-documented faults and moral lapses, recognized its value—sometimes generously, sometimes grudgingly—throughout the more than fifteen centuries of its existence prior to Galileo. The people of the Renaissance thought themselves witnesses to the rekindling of learning in a previously darkened Europe, and thus gave the name Middle Ages to the period between the end of the Classical era and the commencement of their own, when Classical learning was rediscovered and humanism was revived. But that is a misinterpretation of history, based on their ignorance of scientific and humanist progress during the thousand years they dismissed so lightly. Today, we know that the Renaissance was not a uniquely inventive period but rather a particularly brilliant cyclical climaxing of developments in the arts and sciences that had deep roots in medieval history. Renaissance men, too, were standing on the shoulders of their immediate predecessors.

As the long era of Christendom came to a close and the secular society was born, what the Church was anxiously, though perhaps incoherently, trying to do was maintain some semblance of moral balance in the accelerating rush of scientific enterprise. This it felt it could not do in the context of materialist, realist philosophy. In the end, its efforts would be condemned as an endeavor to crush intellectual curiosity. Its attempts to leash what we now call progress were all but fruitless. The hunger for what science had to offer was too overwhelming.

It became evident that the most important result of Galileo's telescopic discoveries had been the rewarding of *curiositas*. What had previously been hidden and apparently inaccessible was revealed, and as a result the morality of self-restraint in the pursuit of scientific knowledge was discredited and quickly abandoned. The long process of the disassociation of usefulness and theoretical efficacy from the good and the true moved on to its culmination in scientism.

From the perspective of the early twenty-first century, it seems clear that while much was gained in the unleashing of science, much was lost as well. For the freedom science offered through theoretical knowledge and material wealth we have paid a heavy price in the loss of any spiritual context for existence and in enslavement to the day-to-day exigencies of technology. We have become commodities, shaped and molded from infancy to meet the requirements of an economy run increasingly on the purest of scientistic, materialist principles. Governments are increasingly powerless to protect their citizens from the excesses of economic demands, because they no longer have access to an unarguable moral authority on which to build the necessary political consensus. And because science has bequeathed to us the unmitigated power to destroy our habitat, we are doing so, rapidly. Aristotle, I think it is safe to say, would have considered the vast majority of us little better than slaves. Socrates, too, would have thought us sorry specimens, mired in base material obsessions.

Questions must be asked. Are we happier in our day-to-day lives than our ancestors, or merely more comfortable? Are the lives we lead more worthy of respect, or less? Is our world, taken all in all, a better place than theirs? To what extent are the advances made over the past four hundred years in social and economic justice attributable to science? In what degree have they been made in spite of science, which teaches the social efficacy of natural selection and survival of the fittest? In seeking answers to questions such as these, in challenging the myths of modernity, we can both write better history and plan a more humane future.

TWENTY-SIX

On the Road to Florence • Galileo's Tomb
Human Happiness • Science and Number
Limits to Scientific Knowledge
An Unfinished Journey

With only a few precious days left before I had to return home from Italy, I was eager to visit Florence to see Galileo's villa at Arcetri and his tomb in the Basilica of Santa Croce. Berkowitz and Sister Celeste were also interested, and the three of us decided to rent a car and travel there together. As it turned out, Florence was something of an anticlimax for us: the villa was closed to the public while undergoing refurbishing as part of the general millennial cleanup under way across Italy, and also, I suspect, as part of the Church's rehabilitation of the scientist it now feels it wronged. But if the destination was somewhat disappointing, the journey, as is so often the case, made the undertaking worthwhile.

Galileo's tomb holds a fascinating story, as befits the resting place of a mythological figure. His will stated his wish to be buried in the family tomb in Santa Croce. When he died, his body was taken to the basilica with a minimum of ceremony and public attention, out of concern for possible interference from Rome. Once he had been safely interred in a small chamber beneath one of the chapels, the question of an appropriate monument within the church was raised. Galileo's supporters wanted something on a grand scale, and they proposed that it be located in the church nave, directly opposite the tomb of Michelangelo. Both Urban VIII and his nephew Cardinal Francesco Barberini warned that this would be inappropriate, given that the scientist had been tried for heresy.

Nevertheless, a campaign to erect an imposing sepulcher continued. There were several reasons for this. The most obvious is that Galileo's supporters genuinely believed his achievements merited recognition comparable to that given Michelangelo, another cultural rebel.* But it was also true that the Medici had no desire to see their considerable investment in Galileo devalued by his remaining under a cloud of suspicion of heresy. They wanted him rehabilitated and buried with appropriate fanfare, to underline the cultural primacy and good government of Florence under the Medici dynasty. And Galileo's scientific supporters, chief among them his student Viviani, feared that continued opprobrium would interfere with the spread of his scientific ideas. Even before his death, Galileo's reputation had been the object of a propaganda war between those in Italy who wanted him remembered as a pious Catholic who had sincerely regretted the "error" of his *Dialogue* and obediently conformed to the ideals of the Counter-Reformation, and those throughout Protestant Europe who were portraying him as a champion and martyr in the battle against an ignorant and degenerate Church of Rome.

The periodic requests for reburial finally received a favorable reply from Rome in 1734, 101 years after the trial, ninety-two after his death. There was an atmosphere of modernization and renewal

---

*It was Viviani, Galileo's chief mythographer, who claimed (incorrectly) that Galileo had been born on the very day of Michelangelo's death, suggesting a transference of spirits. And Galileo's friend the painter Ludovico Cigoli had once written him to commiserate over his Aristotelian "enemies": "I believe the same happened as when Michelangelo began to design buildings outside the orders of the others of his time, when all united to claim that Michelangelo had ruined architecture by taking so many liberties outside of Vitruvius; I replied to them that Michelangelo had ruined not architecture but the architects, because if they lacked designs such as his and continued to work as before, they appeared to be worthless things."

in the Church in the first third of the eighteenth century, and republication of all of Galileo's works, except the *Dialogue*, "Letter to Christina," and other "Copernican" works expressing his scientific-mechanistic views, had been permitted in Florence in 1715. (An earlier edition, also omitting the *Dialogue* and other works, had been published by Viviani in Bologna in 1656.)

This successful solicitation had come from a group of Florentine intellectuals and Freemasons.[1] The Church offered no reasons for giving its permission after so many prior refusals, but the Holy Office did set out some stipulations about the nature of the ceremonies surrounding the reburial and the wording of the epitaph. In neither case, it insisted, was criticism of the 1633 proceedings to be allowed.

It was three years before the tomb had been completed and the transfer of Galileo's remains could take place. The monument was stylistically similar to that of Michelangelo, though it had only two grieving statues beneath the bust of the deceased, as opposed to Michelangelo's three. This was a last-minute alteration: apparently in deference to the sensibilities of the Holy Office, the statue of Philosophy was deleted, leaving only the allegorical images of Mathematics and Astronomy. The true nature of his offense had not been forgotten.

Viviani's notion of a spiritual concordance between Michelangelo and Galileo was given further reinforcement by the reburial ceremony, which was staged March 12, 1737, at 6 P.M., the same day and time as the remains of Michelangelo had been interred in Santa Croce in 1564.

The first order of business was to disinter Viviani, so that he could be buried alongside Galileo as he had wished. This was done, and the casket was opened so that the corpse could be positively identified. A lead plate was found among the bones naming Viviani, and the coffin was closed and placed in the new tomb. Next, the crowd of participants and onlookers moved to the little crypt where Galileo's body lay. Workers broke through a brick wall and extracted a wooden coffin. To the astonishment of everyone, a second wooden casket was found beneath it. Moreover, the lid on the first casket had been smashed, and there were bits of plaster mixed in with the bones. Doctors on hand for the purpose made a careful survey of the remains, but no identifying objects or inscriptions could be found. It was "the corpse of an old man which had once been cut and opened, as was demonstrated by the anatomy professors present," reads the notary's report. The lower jaw had become detached from the skull. The examiners removed three fingers, and a vertebra and a tooth were also removed to be preserved like relics of the saints at the Museum of the History of Science in Florence, the University of Padua, and other locations sacred to science and its principal hero.

Given that no positive identification had been made of the body, curiosity about the contents of the second coffin was at a high pitch. When it was opened and the remains examined, it was with some relief that the doctors announced it was the corpse of a woman. But this coffin, too, bore no objects to help in identifying the body.

After a hurried conference, it was decided to place both coffins in the new tomb, on the assumption that the male must be Galileo, and the female might well be his favorite daughter, Sister Maria Celeste. (Her grave has never been identified: nuns were often buried anonymously.) If the second corpse is indeed that of the daughter, it could well have been placed in Galileo's first resting place by Viviani, though there is no evidence to confirm this—or, for that matter, the identity of the male corpse.

I had agreed to meet Berkowitz and Sister Celeste in front of my hotel, and I sat with my overnight bag on a stone bench, reading the morning's *International Herald Tribune*.

I had been waiting there in the shimmering morning heat for several minutes when a silver S-series Mercedes rolled up, its fat tires on the hot pavement making a sound of tearing silk. The tinted passenger-side window slid silently down to reveal a beaming Sister Celeste, her lovely face framed in white linen. There was a little gust of cool leather-scented air. I stooped to look inside, and there was Berkowitz behind the wheel, grinning from ear to ear.

"Where did you find this?" I asked him.

"Hertz," he said. "I thought, What the hell, if we're taking a trip, we should do it right. You only live once—right, Sister?"

The trunk lid had popped up at the push of a button somewhere, and I walked to the rear of the car to put my bag inside. It was a cavernous space. There were two nondescript but well-made navy blue nylon bags and a large suitcase on one side of the carpeted floor. A sealed case of Prosecco Spumanti held pride of place in the center. I put my own battered bag beside it and closed the lid.

Our excursion to Florence was turning into something more akin to an expedition. Berkowitz had begged to be allowed to plan the whole thing, and what's more, he insisted on paying for our food and accommodation.

"What's money for?" he'd explained.

Exactly where we'd be staying was to be a surprise, but the trip was to take roughly as long as it had in Galileo's time—about four days.

I slid into the back seat, pulled the door closed behind me (it made a satisfying "thunk"), and we were off, though there was no sensory evidence of a means of propulsion. It was like riding on a flying carpet. We were soon on the city's ring road and from there we exited north on a four-lane highway.

"First stop is Borgo Paraelios. I thought we'd get some lunch there. It's just forty or fifty kliks north of the city. It's a pretty fabulous place—a nineteenth-century family villa. You get the run of the house if you're staying there; original art everywhere; a good library, if you read Italian. Indoor and outdoor pools. And there's only about a dozen guest rooms. It's the next best thing to knowing an Italian countess, which I unfortunately don't."

Berkowitz poked a button and the radio came to life with the glorious *Dies Irae* movement of the Verdi Requiem. "The volume increases with the speed," he said.

"What?"

"The volume of the radio. It goes up automatically as the speed of the car increases. You don't have to turn the knob."

"Now there's a technological feat," I said.

"Remarkable," said Sister Celeste.

Berkowitz chuckled good-naturedly. "All right, you guys. So I'm simple-minded. But Galileo did this trip on a mule, with the plague raging around him. Which do you prefer?" He was holding his right hand above the steering wheel, palm up. He flipped his palm down, then back upward. "Mule or Mercedes? Mercedes or mule? Which is it?"

"With all due respect, Mr. Berkowitz," the little nun said, "that is a meaningless question."

"Meaningless?"

"Well, perhaps not entirely meaningless, but certainly irrelevant. The relevant question is one of human happiness. Were individual humans more or less happy in Galileo's time than they are now? That is the only question worth asking. There is no point in asking whether they are better off materially, because quite obviously they are."

"And what is your answer to the question of happiness, Sister?" I asked her.

"You know, as much as I've thought about it, I'm still not entirely sure. So many modern histories are written from a point of view that takes for granted the equation between material well-being and happiness. But if you read the authors writing of their own times, it does not seem to me that they were any more or less happy than we are. Perhaps the answer is that given a certain basic level of security, material wealth is irrelevant to happiness. I'm afraid that's the best I'm able to do."

She turned to Berkowitz. "And perhaps Galileo, while he was not able to listen to beautiful music on his mule, could hear the birds singing and the wind in the leaves and smell new-cut hay and wildflowers."

"I doubt that he'd have turned down an even swap, Mercedes for mule," Berkowitz insisted.

"Perhaps you're right," Sister Celeste replied. "But you must remember that the car would have been useless to him without the vast infrastructure we've built to go along with it, to make it work. Oil exploration, supertankers, pipelines, air pollution, paved streets and expressways, highways, parking lots, gas stations, steel mills, rubber plantations, chemical factories, junkyards, urban sprawl, supermarkets, malls, uninhabitable city centers, death, and mutilation on a truly tragic scale and so on." She laughed. "Have I missed anything? When you accept a technology, Mr. Berkowitz, you are also accepting all its ancillaries. And it is usually the ancillaries that cause damage. If you were to have shown him what in reality he would be taking on by accepting your automobile, he might well have refused, do you not think?"

"Or take television . . ." I said.

"*You* take television," Berkowitz said. "Even I wouldn't be fool enough to trade a mule for it."

It was his grudging way of granting the validity of the nun's point without actually saying so.

We were silent for a moment, listening to the oboe and mezzo-soprano in a lovely, soaring duet.

"Why don't I get us off this highway and on to some country roads where we can roll down the windows and enjoy the scenery," Berkowitz said at last.

This was not as good an idea as it first seemed to be. Several tens of thousands of other Italian motorists appeared to have had the same plan, as well as a good number of truck drivers. Our progress slowed to a crawl, and we had to leave the windows up to keep the dust and fumes out. Still, it was exciting to know that we were following the route of the ancient Via Salaria, the road connecting the salt pans at the Tiber's mouth with the Etruscan settlements around present-day Rome and beyond, in the Sabine Hills.

I tried to summarize for Berkowitz and the nun an idea I'd been making notes on for the past several weeks. It began with the observation that today's science has come to the point where it insists on three things. The first of these is that everything that goes on in nature can be accounted for by matter and motion. This is exactly what Democritus and the Greek mechanists had said, and it's what Aristotle was reacting against with his vitalism and teleology. Twenty-four hundred years later, scientists don't talk about atoms bouncing off one another—they're more sophisticated than that. Now they say that everything in nature can be accounted for by the movement of photons, the movement of electrons, and the interaction between photons and electrons.[2]

"The second thing," I continued, "is that modern science insists, just as Galileo did, that reality consists of numbers and formulas. Galileo said philosophy is written in the great book called the universe, and that the language it is written in is the language of mathematics. Give me a second here and I can give you the modern version of that."

I rummaged in my shoulder bag and found my tattered copy of John Gribbin's *Schrödinger's Kittens*.

"Let me read you this. This is coming from an astrophysicist who is also one of the top science writers in the English language.

He's talking about James Clerk Maxwell's equations describing electromagnetic field theory, which is a cornerstone of modern quantum physics. In fact, a lot of scientists think that the 'field' the equations describe is the fundamental essence of reality. Here's what Gribbin says.

> In the modern interpretation of Maxwell's equations, the ether and the vortices [of earlier theorists] have vanished, and been replaced by the reality of . . . lines of force, the electromagnetic field. Of course, this is only the latest mental image to hold center stage; we have no better idea of what is "real" for an electron than Faraday, or Maxwell, or anyone else had. The advantage of the field theory is that it is simple, and that it shows the way the mathematics works in a clear-cut way. But models should never be regarded as anything more than an aid to the imagination, a way of helping us to picture (or calculate) what is going on. *The reality resides in the mathematical equations themselves*, whether the equations are used to describe electromagnetic waves, heat in a solid, or the flow of water. As long as the equations correctly tell us how the system will change when it is disturbed in a certain way, it doesn't really matter how you picture the interplay of forces at work [emphasis added]."[3]

Sister Celeste had been listening intently. "He is really saying two very interesting things. He is saying first of all, with Galileo, that the story of philosophy, or knowledge of the world, is written in mathematics—in other words, mathematical relationships constitute the underlying reality of the world.* Then he goes on to say that one is free to visualize or model what goes on in any way, as long as it doesn't contradict fundamental mathematics."

"That's right: '. . . it doesn't really matter how you picture the interplay of forces at work.'"

---

* A less rigorous mathematical realist might say instead that quantifiable properties, expressed mathematically, constitute the underlying reality of the world. (This implies the existence of properties beyond the mathematical expressions.) This does not, however, seem to have been the position of Galileo, and is a weaker form of mathematical realism than that expressed here by Gribbin.

"I find that most interesting," she went on, "because it is not so very different from what Cardinal Bellarmine told Galileo in 1611. He said, 'Feel free to build new models to explain what we observe in nature, but be careful not to lose sight of the fact that they are just models, and that the reality they mimic is knowledge of another order.' Science, as your Mr. Gribbin says, has at least progressed beyond Galileo to the point that it now accepts that it is possible for two or more different models to adequately account for the same observed phenomena. I suppose it could hardly deny that point, having by now lived through several revolutions, each of which put in place a new paradigm. I suppose that's become simply a fact of history.

"Where Mr. Gribbin, and the less articulate scientists he speaks for, differ from Bellarmine is that he sees the underlying reality of nature as being mathematical formulas, whereas Bellarmine saw it as something beyond the capacity of human reason to grasp, at least in our present state."

"Well, that's progress," Berkowitz commented.

"Is it? I'm not so sure. If we accept that reality is mathematical formulas, what we are saying is that reality is a thing made by human reason. Because mathematics is a human invention, yes? Bellarmine would have gone only halfway in accepting this, I think. He would have agreed that the world as *we experience it* is something created by our minds out of a preexisting potential existing 'out there' beyond ourselves. That's a basic Platonic and Aristotelian idea. It is that preexisting potential, what Aristotle called 'matter *in potentia*' that is the 'real' reality. And what, precisely, that is, nobody knows for sure."

"But, Sister," I said, "you've missed something. I don't think it's true any longer to say that science doesn't acknowledge any reality beyond its own mathematical formulas. In fact, I don't think it's been true since about 1920 and the Copenhagen Interpretation of quantum physics, which is still the gold standard."

"I'm supposed to know what the hell the Copenhagen Formula is?" Berkowitz interjected. It was a complaint rather than a question.

"The Copenhagen *Interpretation*. All you need to know is that it's the orthodox dogma of modern physics, and it says that quanta, the subatomic entities that make up the world, are nothing more or less than thickenings or concentrations in a universal electromagnetic force field. Quanta have the peculiar characteristic of occurring as something called 'probability waves' until they're observed by a conscious observer. When that happens, the probability wave collapses and the quantum entity either exists or does not exist. Until the observer arrives on the scene, it exists only *in potentia*, as Aristotle might have said if he'd been around in the 1920s. This is something that has been verified in experiment after experiment. With me so far?"

"Carry on . . ."

"Okay. Consider this. The observer is also made up of quanta, which have no existence until observed. Right?"

"Right . . ."

"So the observer must be observed himself before he can do any observing, because until he is observed, he doesn't exist. Okay?"

"Okay, I think I see where you're going . . ."

"It's not me! This is orthodox science, and has been for about eighty years! You can read it in any number of texts, including Stephen Hawking's *A Brief History of Time*."

"That book is unreadable. I've tried."

"All right, but that's beside the point. The point is that modern physics has this huge unresolved problem that it just sweeps under the carpet because the system works fine on a practical, day-to-day instrumental level without having to solve it. The problem is, where does the regression of observers, and observers of observers, and observers of observers of observers, and so on . . . where does it end? It can't be an infinite regress—that would be absurd. So somewhere the buck has to stop, with an ultimate observer. There's no way out of it."

"The G word."

"Exactly. Or something like it. So when science claims that the ultimate reality is in the mathematical formulas, it has its fingers crossed behind its back. It knows better. It *knows* that there must be a reality behind scientific reality."

Sister Celeste made some flattering comments on my analysis, then addressed me directly.

"It occurs to me now that there is another way that science knows its numbers cannot possibly be the ultimate reality, no matter what Mr. Gribbin or his colleagues say. It is really quite simple, and it arises out of Heisenberg's Uncertainty Principle."

She turned to Berkowitz. "You are familiar with this?"

"I think so," he said. "The principle says that you can't know about both the mass and the speed of a quantum particle at the same time. In other words, if you know the mass, you can't know the speed, and vice versa. It's not just difficult; it's impossible in principle. The formulas won't allow it."

"That will do nicely, although to be perfectly orthodox we should be saying *position* and *momentum* instead of speed and mass. But it really makes no difference for my point."

She paused for a moment to collect her thoughts. "And my point is this. Science claims that you must account for everything that goes on in nature by reference to matter and motion, by the movements of photons and electrons and their interactions. But the Uncertainty Principle says you can't know about both matter and motion at the same time. It seems to me that science is once again saying that it cannot know reality. And not because it's difficult, as you correctly observed, Mr. Berkowitz, but because it is impossible in principle."

"So what you two are saying is that modern science has proved to its own satisfaction that it is incapable of describing reality," Berkowitz remarked.

Sister Celeste and I looked at each other and nodded.

"That's right," I said. "Science cannot describe reality, for reasons it figured out itself. But it also figured out that there must nevertheless *be* a reality beyond its mathematical models. Which means that in the Galileo affair, which was all about the conflicting scientistic and religious interpretations of reality, the Church was right and Galileo was wrong. QED, as Feynman might say."

"And that's it? I mean, that's the final conclusion?"

"That's it," I said, with some relief.*

---

*There is another, ancient argument that is, in my opinion, irrefutable in logic. Suppose a collection of scientific hypotheses were able to explain all known appearances. What can be concluded is that they *may* be true, not that they are *necessarily* true. In order for them to be necessarily true, it would have to be proved that no other system of hypotheses could possibly be imagined that could explain the appearances just as well. (This is of course closely related to the argument proposed by Cardinal Bellarmine and Urban VIII.)

Berkowitz seemed singularly unimpressed. "But I've got two more days all booked and arranged before we get to Florence!"

"That's okay. In fact, that's great! We can all talk about something else for a change."

"For instance?"

"For instance, how we now know more certainly than ever before in history—because science tells us so—that there is a reality beyond what sensory experience and reason can show us, that there is a reality that transcends science."

"And?"

I was becoming a little impatient. "And don't you think that's worth knowing, after four hundred years of being told 'if you can't measure it, it doesn't exist'? Don't you think science is overdue for a course in humility?"

"What excites me," Sister Celeste said, "is the prospect of seriously looking for ways to link the two levels of reality once again. That would mean incorporating nonquantifiable considerations such as goodness and morality into physical science. It would mean building a teleology, a moral goal for science, so that it could begin to consider where it is going, and where it is taking the rest of us. We have the opportunity to reconnect moral and natural science in a way that would be beneficial to modern society."

"I don't know," said Berkowitz, "it sounds like a pipe dream to me."

The little nun shot back, "The pipe dream, Mr. Berkowitz, is one-dimensional scientistic reality. It cannot last, because it is not true. We might as well be thinking of how to give it a decent burial. It has been a minor diversion, a dead-end channel in the river of human history that flows from the earliest civilizations of the Indus Valley through to the present. A singularly perverse diversion, at that, I would venture to say."

At that point we reached the toll station at Tarquinia and diverted ourselves with a search for the correct change. Once through the gates, we pulled over onto the shoulder to stretch our legs. The air was still, and the heat and humidity were oppressive, as we stood on asphalt so close to the sea.

APPENDIX

On the Question of Galileo's "Perjury"

Five months after the death of his daughter Sister Maria Celeste in 1633 Galileo—grief-stricken, ill, his eyesight failing—wrote a despondent, much-quoted letter to his French supporter Nicolas Peiresc. In it he strangely echoes his fellow Tuscan Niccolò Machiavelli, whose famous book of politics, *The Prince*, had been published 120 years earlier and dedicated to Lorenzo de' Medici, the illustrious ancestor of Galileo's patron: "I do not hope for any relief, and that is because I have committed no crime. I might hope for and obtain pardon, if I had erred; for it is to faults that the prince can bring indulgence, whereas against one wrongfully sentenced while he was innocent, it is expedient, in order to put up a show of strict lawfulness, to uphold rigor. . . ." Galileo goes on to suggest that the prohibition issued in 1616 had resulted from deceit, before making a cryptic reference to the motives of his enemies:

> But how clearly would appear my most holy intention if some power would bring to light the slanders, frauds, stratagems, and trickeries that were used eighteen years ago in Rome in order to deceive the authorities! . . . You have read my writings, and from them you have certainly understood which was the true and real motive that caused, under the lying mask of religion, this war against me that continually restrains and undercuts me in all directions, so that neither can help come to me from outside nor can I go forth and defend myself, there

> having been issued an express order to all Inquisitors that they should not allow any of my works to be reprinted which had been printed many years ago or grant permission to any new work that I would print . . . so that it is left to me only to succumb in silence under the flood of attacks, exposures, derision, and insult coming from all sides.[1]

Those who would argue that Galileo's abjuration was sincere have this letter to contend with. Its bald statement, "I have committed no crime," has traditionally been taken as evidence for the contention that Galileo's abject confession of guilt before the Inquisition was willfully perjured or extorted under threat of torture.

But this has to be weighed in the context of another passage from the same letter, which is less frequently quoted: "I have two sources of perpetual comfort, first that in my writings there cannot be found the faintest shadow of irreverence towards the Holy Church; and second, the testimony of my own conscience, which only I and God in Heaven thoroughly know. And He knows that in this cause for which I suffer, though many might have spoken with more learning, none, not even the ancient Fathers, have spoken with more piety or with greater zeal for the Church than I."[2]

Taken together, these quotations give a different meaning to the letter. What Galileo seems to be saying is that he has "committed no crime" in pursuing his project of trying to convert the Church to the new philosophy of mathematical realism—"this cause for which I suffer." It is his dedication to this same cause that is "the true and real motive that caused, under the lying mask of religion, this war against me."

But the philosophical issue of Galileo's realism versus the Church's anti-realism did not arise at his trial, which confined itself to the fact of his having disobeyed the 1616 injunction. It was this disobedience, and this alone, that he confessed and abjured. It is the more serious crime of impiety, of trying to undermine the authority of the Church, of which Galileo is declaring his innocence to Peiresc. His motives, he insists, were pure, and he had acted in the best interests of the Church.

This is certainly the most straightforward interpretation of the text, because it avoids both the necessity of conjuring an elaborate and vindictive conspiracy against Galileo within the Church, along with the necessary motive and actors, and the assumption that Galileo committed perjury. That Galileo would lie under oath seems, from everything we know about the man, highly improbable. And the notion that there was a conspiracy against him is to be found mainly in his own letters, and those of admirers reacting to his suspicions and accusations. There is little or no documentary evidence of conspiracy—on the contrary, we know that throughout his career he manufactured "enemies" to suit his rhetorical purposes.

Finally, if the motive imputed by Galileo to his enemies, real or manufactured, is not their wish to derail his philosophical project, what was it? The only possible answer is personal vendetta, and this is the line taken in most histories of the Galileo affair. Usually it is the Jesuit order that is blamed, although sometimes specific persons are named as the malevolent mastermind behind events.[3]

Aside from making the interpretation of the historical record needlessly complicated, this view ignores the fact that the Inquisition process was carefully contrived to weed out exactly this kind of indictment, and it imposed very serious penalties on anyone caught making false accusations. Had Galileo and his supporters seriously believed that a conspiracy was afoot and that he had been falsely accused, one would have expected Galileo or someone else to say so to the Holy Office. But there is no record of any such charge. Once again the documentary evidence favors the simplest explanation.

Further evidence for Galileo's sincerity in his abjuration is contained in a much later letter to Francesco Rinuccini (1641) in which Galileo appears to have accepted the arguments put forward by Bellarmine and Urban VIII against his mathematical realism:

> To be sure, the conjectures by which Copernicus maintained that the Earth is not at the center are all removed by that most sound argument, taken from the omnipotence of God. He being able to do in many, or rather, infinite ways, that which to our view and observation seems to be done in one particular way, we must not pretend to hamper God's hand and tenaciously maintain that in which we may be mistaken. And, just as I deem inadequate the Copernican observations and [its] conjectures to be insufficient, so I judge equally and more,

fallacious and erroneous those of Ptolemy, Aristotle, and their followers when without going beyond the bounds of human reasoning their inconclusiveness can be very easily discovered.[4]

NOTES

ONE

1. Maurice Finocchiaro, *The Galileo Affair: A Documentary History*, "Introduction" (University of California Press, 1997), 5.

TWO

1. Quoted in Giorgio de Santillana, *The Crime of Galileo* (Time Inc., 1962), 144.
2. Ibid.
3. Robert S. Westman, "The Copernicans and the Churches," in *God and Nature: Historical Essays on the Encounter between Christianity and Science* (University of California Press, 1986), 86.

FOUR

1. Hans Blumenberg, *The Legitimacy of the Modern Age*, trans. Robert M. Wallace (MIT Press, 1985), 390.
2. Johannes Kepler, Letter to Maestlin, 2 August 1595, *Werke* 13:27, quoted in Richard S. Westfall, "The Rise of Science and the Decline of Orthodox Christianity," in David C. Lindberg and Ronald L. Numbers, eds., *God and Nature: Historical Essays on the Encounter between Christianity and Science* (University of California Press, 1986), 221.
3. Kepler, *The Six-Cornered Snowflake*, trans. Colin Hardie (Clarendon Press, 1966), 33.
4. Kepler, *Epitome of Copernican Astronomy*, in *Great Books of the Western World*, ed. Robert M. Hutches, vol. 16 (Encyclopaedia Britannica, 1952), 916–17.
5. Kepler, *Epitome of Copernican Astronomy*, quoted in Westfall, "The Rise of Science and the Decline of Orthodox Christianity," in *God and Nature*, 223.

6. Kepler, Letter to Maestlin, 15 April 1597.
7. This and subsequent quotations in this paragraph are from Arthur Koestler, *The Sleepwalkers* (Penguin/Arkana, 1989), 245.
8. Quoted in Koestler, *The Sleepwalkers*, 245.

FIVE

1. Ludovico Geymonat, *Galileo Galilei: A Biography and Inquiry into His Philosophy of Science* (McGraw-Hill, 1965), 9.
2. Dudley Shapere, *Galileo: A Philosophical Study* (University of Chicago Press, 1974), 74ff.
3. Quoted in Erwin Panofsky, "Galileo as a Critic of the Arts" (Martinus Nijhof, 1954), 9.
4. Ibid.
5. Charles Singer, *A Short History of Scientific Ideas to 1900* (Clarendon Press, 1959), 24.

SIX

1. Nicolaus Copernicus, Preface to *De Revolutionibus*, quoted in Thomas S. Kuhn, *The Copernican Revolution: Planetary Astronomy in the Development of Western Thought* (Harvard University Press, 1985), 137.
2. Copernicus, *De Revolutionibus*, in Kuhn, *The Copernican Revolution*, 151–52.
3. *International Herald Tribune*, Italy Edition, 9 July 1999.

SEVEN

1. For a full discussion of these distinctions, see Dorothea Krook, *Three Traditions of Moral Thought* (Cambridge University Press, 1959), esp. Part 1.
2. Giorgio de Santillana, *The Origins of Scientific Thought* (Mentor, 1961), 213.
3. Santillana, Introduction to Galileo Galilei, *Dialogue on the Two Chief World Systems*, trans. T. Salusbury (University of Chicago Press, 1955), 6.
4. W. G. de Burgh, *The Legacy of the Ancient World* (Penguin, 1967), 211–12.

EIGHT

1. Phillip Frank, *Einstein: His Life and Times* (Alfred A. Knopf, 1947), 7–8.
2. Robert Jastrow, *God and the Astronomers* (W.W. Norton, 1978).
3. See, for example, Robert S. Westman, "The Copernicans and the Churches," in *God and Nature: Historical Essays on the Encounter between Christianity and Science* (University of California Press, 1986), 76ff.

NINE

1. Quoted in Kuhn, *The Copernican Revolution*, 186.
2. Quoted in Koestler, *The Sleepwalkers*, 361.
3. Ibid., 364.
4. Galileo Galilei, *The Starry Messenger*, in Stillman Drake, *Discoveries and Opinions of Galileo* (Anchor Books, 1957), 57.
5. Ibid., 49.
6. See, for example, Santillana's *Crime of Galileo*, which quotes as evidence for the supposed fatuousness of Galileo's critics Galileo's own highly colored paraphrases in *Dialogues* (Chapter 1).
7. Quoted in Drake, *Discoveries and Opinions of Galileo*, 73.
8. Quoted in Drake, *Discoveries and Opinions of Galileo*, 37.
9. Geymonat, *Galileo Galilei*, 85.
10. Santillana, *The Crime of Galileo* (University of Chicago Press, 1955).
11. Quoted in Silvio A. Bedini, "Galileo and Scientific Instrumentation," in William A. Wallace, ed., *Reinterpreting Galileo* (The Catholic University of America Press, 1986), 138.
12. Quoted in Drake, *Discoveries and Opinions of Galileo*, 65.
13. Quoted in Santillana, *The Crime of Galileo*, 4. A more complete translation is to be found in Drake, *Discoveries and Opinions of Galileo*, 67–68.
14. Ignatius Loyola, *The Spiritual Exercises*, trans. Anthony Mottola (Doubleday, 1964), 140–41.

TEN

1. Drake, *Discoveries and Opinions of Galileo*, 60.
2. Ibid., 61.
3. Galileo Galilei, *Letters on Sunspots*, in Drake, *Discoveries and Opinions of Galileo*, 143.
4. Ibid., 82.
5. James M. Lattis, *Between Copernicus and Galileo* (University of Chicago Press, 1994), 184ff.
6. Quoted in Lattis, *Between Copernicus and Galileo*, 190.
7. Ibid., 188.
8. Ibid., 183.
9. Ibid.

ELEVEN

1. Josef Pieper, *Guide to Thomas Aquinas* (Mentor, 1964), 47.
2. Quoted in de Burgh, *The Legacy of the Ancient World*, 455.
3. Thomas Aquinas, *Summa Theologiae* II 2, q.166, a.2, in *St. Thomas Aquinas, Summa Theologiae*, trans. Thomas Gilly (McGraw-Hill, 1971) vol. 44, 199.

4. Aquinas, *In Aristotelis libros de caelo expositio* II lect.7, n.4 (364), quoted in Blumenberg, *The Legitimacy of the Modern Age*, 333.
5. Ibid., 334.
6. Aquinas, *de Malo*, q.11, a.4, quoted in Blumenberg, *The Legitimacy of the Modern Age*, 335.
7. In *Reinterpreting Galileo*, ed. Wallace, 85.

TWELVE

1. Drake, *Discoveries and Opinions of Galileo*, 81.
2. Letter to Paolo Gualdo, May 1612, quoted in Santillana, *The Crime of Galileo*, 12.
3. Quoted in Drake, *Discoveries and Opinions of Galileo*, 151–52.
4. John H. Taylor, *The Literal Meaning of Genesis,* vol. 1 (Newman, 1982), 42–43. Quoted by Galileo in the "Letter to the Grand Duchess Christina."
5. Galilei, *Dialogue on the Two Chief World Systems*, ed. Santillana, 48.
6. Galilei, "Letter to the Grand Duchess Christina," in Drake, *Discoveries and Opinions of Galileo*, 210.
7. Ibid., 183.
8. St. Jerome, Letter 53 to Paulinus, EN V 323; *GA*, 99, quoted in "Galileo on Science and Scripture," in *The Cambridge Companion to Galileo*, 289.
9. Quoted in Drake, *Discoveries and Opinions of Galileo*, 157.
10. Ibid., 163–64.

THIRTEEN

1. Quoted in Santillana, *The Crime of Galileo*, 117.
2. Quoted in Drake, *Discoveries and Opinions of Galileo*, 158–59.
3. Ibid., 118.
4. Ibid., 119–20.
5. Ibid., 124.
6. Ibid., 126.
7. Santillana, *The Crime of Galileo*, 135–36, 2.
8. René Descartes, *Principles of Philosophy*, in Elizabeth S. Haldane and G. R. T. Ross, eds. *Philosophical Works of Descartes*, vol. 1 (Cambridge University Press, 1967), 203ff.
9. Quoted in Michael John Gorman, "A Matter of Faith: Christopher Scheiner, Jesuit Censorship, and the Trial of Galileo," *Perspectives on Science* 4, no. 3 (1996).
10. Ibid., 131.
11. Quoted in Santillana, *The Crime of Galileo*, 140.

FOURTEEN

1. Bertrand Russell, *History of Western Philosophy* (George Allen & Unwin Ltd., 1965), 782.

FIFTEEN

1. See Blumenberg, *The Legitimacy of the Modern Age*, 138ff.

EIGHTEEN

1. *The Assayer*, in Drake, *Discoveries and Opinions of Galileo*, 277.
2. Ibid., 241.
3. Galilei, *Il Saggiatore (The Assayer)*, quoted in *The Cambridge Companion to Galileo*, 53.
4. Galilei, *Il Saggiatore (The Assayer)*, quoted in Drake, *Discoveries and Opinions of Galileo*, 271.

NINETEEN

1. Quoted in Santillana, *The Crime of Galileo*, 183.
2. See, for example, W. R. Shea, *Galileo's Intellectual Revolution* (Science History Publications, 1972), 172–89.
3. See Finocchiaro, *The Galileo Affair: A Documentary History* (1989), 33.
4. Galilei, *Dialogue on the Two Chief World Systems*, ed. Santillana, 129.
5. Ibid., 131.
6. Santillana, *The Crime of Galileo*, 187.
7. The letters are excerpted from Geymonat, *Galileo Galilei.*
8. See Gorman, "A Matter of Faith: Christopher Scheiner, Jesuit Censorship and the Trial of Galileo," for the best of recent examinations of this issue.
9. See, for example, Santillana, *The Crime of Galileo*, esp. chapter 13.
10. Quoted in ibid., 204–5.
11. Quoted in Geymonat, 140.
12. Ibid., 229.

TWENTY

1. Quoted in Charles Blitzer, *Age of Kings* (Time Inc., 1967), 78.
2. Ibid.

TWENTY-ONE

1. Friedrich Heer, *The Holy Roman Empire* (Phoenix, 1968), chapter 10.
2. Sir George Norman Clark, *The Seventeenth Century* (Oxford University

Press, 2nd ed., 1961), 277.
3. Ibid.

TWENTY-TWO

1. Rousseau to Voltaire, 10 September 1755, quoted in Blumenberg, *The Legitimacy of the Modern Age*, 494.
2. Quotations are taken from the English text published in *L'Osservatore Romano* N. 44 (1246), 4 November 1992, available at www.cco.calatech.edu/-newman/sci-0211.html
3. This line of thought was first suggested to me by William A. Wallace, *Galileo, the Jesuits and the Medieval Aristotle* (Variorum, 1991), chapter 4.

TWENTY-THREE

1. This and subsequent quotations from Niccolini's letters are taken from Santillana, *The Crime of Galileo*, chapter 10.
2. Quoted in Shea, "Galileo and the Church," 235.
3. Aquinas, *Summa Theologiae* (ii-ii, qu.64 art. 2–5.), quoted in Henry Charles Lea, *The Inquisition of the Middle Ages: Its Organization and Operation* (Eyre and Spottiswoode, 1963).
4. See Immanuel Le Roy Ladurie, *Montaillou: The Promised Land of Error* (Vintage Books, 1979). Quotations from Inquisitional archives throughout are illustrative.
5. Information on the Inquisition is compiled from various sources: Jean Guiraud, *The Medieval Inquisition*, trans. E. C. Messenger (Burns, Oates and Washbourne Ltd., 1929); Henry Charles Lea, *The Inquisition of the Middle Ages: Its Organization and Operation* (Eyre and Spottiswoode, 1963); Jacques Madaule, *The Albigensian Crusade*, trans. Barbara Nail (Burns and Oates, 1967); Hoffman Nickerson, *The Inquisition: A Political and Military Study of Its Establishment* (John Bale and Sons, and Danielson, 1929); James B. Given, *Inquisition and Medieval Society: Power, Discipline and Resistance in Languedoc* (Cornell University Press, 1997); *The Encyclopaedia Britannica* (eds. 11 and 15).
6. Santillana, *The Crime of Galileo*, 260.
7. Ibid., 267.
8. Ibid., 272–73.
9. Ibid., 273.
10. Ibid., 280.
11. Ibid., 277, n11.
12. Ibid., 327.

TWENTY-FOUR

1. Quotations from the trial transcripts are taken from Santillana, *The Crime of Galileo*, chapter 15.

TWENTY-FIVE

1. Santillana, *The Crime of Galileo*, 5–6.
2. Ibid., 421.
3. Ibid., 397.
4. John Milton, *Areopagitica* (Harvard Classics, P.F. Collier & Son, 1937), 189.
5. Jan de Vires, *The Economy of Europe in an Age of Crisis, 1600–1750* (Cambridge University Press, 1976), 27ff.

TWENTY-SIX

1. This account follows that of Paolo Galluzzi, "The Sepulchers of Galileo," trans. Michael John Gorman, *The Cambridge Companion to Galileo*, 417ff.
2. See, for example, Richard Feynman, *QED: The Strange Theory of Light and Matter* (Penguin, 1990), 110.
3. John Gribbin, *Schrödinger's Kittens and the Search for Reality* (Little, Brown and Co., 1995), 67.

APPENDIX

1. Gribbin, *Schrödinger's Kittens and the Search for Reality*, 351.
2. Mary Allan-Olney, *Private Life of Galileo, Compiled Principally from His Correspondence and That of His Eldest Daughter, Sister Maria Celeste* (Macmillan, 1870), 278–79.
3. Chief among these has been Father Chistopher Scheiner. Newly uncovered letters now seem to rule him out. See Gorman, "A Matter of Faith?"
4. Stillman Drake, *Galileo at Work: His Scientific Biography* (University of Chicago Press, 1978), 417.

INDEX